Natural Stone and World Heritage

T0174984

Natural Stone and World Heritage
Series editor: Dolores Pereira

Natural Stone and World Heritage
Salamanca (Spain)
Dolores Pereira

Natural Stone and World Heritage
Delhi-Agra, India
Gurmeet Kaur, Sakoon N. Singh, Anuvinder Ahuja, and Noor Dasmesh Singh

Natural Stone and World Heritage
UNESCO Sites in Germany
Edited by Angela Ehling, Friedrich Häfner, and Heiner Siedel

Natural Stone and World Heritage
The Castles and Town Walls of King Edward in Gwynedd
Ruth Siddall

For more information about this series, please visit: www.routledge.com/Natural-Stone-and-World-Heritage/book-series/NSWH

Natural Stone and World Heritage
The Castles and Town Walls of King Edward in Gwynedd

Ruth Siddall
University College London
London, UK

 CRC Press
Taylor & Francis Group
Boca Raton   London   New York   Leiden

CRC Press is an imprint of the
Taylor & Francis Group, an **informa** business

A BALKEMA BOOK

First published 2022
by CRC Press/Balkema
Schipholweg 107C, 2316 XC Leiden,
The Netherlands
e-mail: enquiries@taylorandfrancis.com
www.routledge.com – www.taylorandfrancis.com

CRC Press/Balkema is an imprint of the Taylor &
Francis Group, an informa business

© 2022 Taylor & Francis Group, LLC

Library of Congress Cataloging-in-Publication Data
A catalog record has been requested for this book

ISBN: 978-0-367-43315-4 (hbk)
ISBN: 978-1-003-00244-4 (ebk)

DOI: 10.1201/9781003002444

Typeset in Times New Roman
by codeMantra

Dedication

This book is dedicated to my mother, Anne Siddall (1940–2020)

Contents

Preface

This book is about the stone used to construct the World Heritage Site (WHS) that encompasses the Castles of Edward I in Gwynedd, but it is also about the overall use of stone throughout human occupation of the region of the castles' hinterlands in the country, primarily in the areas of eastern Anglesey, the location of Beaumaris Castle, the coastal region of Arfon which includes Caernarfon and its Castle and the cathedral city of Bangor, and the regions surrounding the castle towns of Conwy and Harlech. It is a geology guide to the Anthropocene of North West Wales. Although it is only the four castles and two town wall circuits that are technically included in the WHS designation, the story of their building materials, using both recycled and freshly quarried stone, cannot be regarded in isolation from the wider use of stone throughout this area and indeed from the regional geology. Stone has been an essential material throughout human history, and archaeologically, stone artefacts represent some of the earliest evidence of people and societies. Despite this, globally, there is an ignorance and one might say even a disinterest of the relationship of geology to the built environment and the importance that appropriately chosen stone contributes to the longevity and character of a building or indeed an urban environment (Kaur et al., 2020). This was the case even at the time of the building of Edward's castles. In the Exchequer accounts which provide inventories of expenditure during the construction of the castles, detail is given about using timber from named woodlands or even from dismantled documented buildings; yet stone is described as simply being *brought to the site*. Its provenance is rarely mentioned, and to find a named quarry in these otherwise highly detailed documents is a rare thing.

The UNESCO WHS of the Castles and Town Walls of King Edward I in Gwynedd is one of just over thirty currently ratified World Heritage Sites located in the British Isles, and it is the one that arguably covers the largest area, extending from Harlech in the south on Cardigan Bay across Snowdonia to Conwy on the north coast, a region that encompasses some 875 km^2 of complex regional geology and a six millennia history of the exploitation of stone. It includes a designated Global Heritage Stone Resource (GHSR) in the form of the Cambrian-aged slates, which have been quarried since the Roman Period. Designated in 1986, the Castles and Town Walls of King Edward I in Gwynedd WHS are, strictly speaking, the four main castles planned and built during Edward I's campaigns and invasion of Wales in the late 13th and early 14th centuries, each one a supreme example of medieval military architecture. They are Caernarfon Castle, Conwy Castle, Harlech Castle and Beaumaris Castle. The castles at Caernarfon and Conwy are connected to walled towns, both with an almost complete and original circuit of walls, which are also included in the WHS designation. These structures fulfil three of UNESCO WHS' ten criteria in that they 'represent a masterpiece of human creative genius', they 'bear a unique or at least exceptional testimony to a cultural tradition or to a civilization which is living or which has disappeared' and finally are 'outstanding examples of a type of building, architectural or technological ensemble or landscape which illustrates significant stages in human history' (UNESCO, 2019).

A disregard for the identification of stone and the connection of these materials to local or regional landscapes is often absent from many reports, descriptions and architectural studies of the built environment, much to the frustration of geologists, geotourists and others who have an interest in sourcing or conserving stone. The statement 'this building is built of the local stone' is used all too frequently, much to the frustration of geologists and with scant regard for what this actually means or how it is applied to the complexity and diversity of local geology and potential quarry sites. This statement is rarely qualified by a summary description of the 'local stone'. I am glad to say that this attitude is changing, very much in part due to the efforts of the International Union of Geological Sciences Heritage Stone Sub-Commission and Heritage Stone Resource (IUGS HSS and IUGS HSR), both working groups within the International Commission on Geoheritage (ICG). These groups are working to promote the importance and knowledge of natural stone in built heritage and identify GSHRs which are building-ing stones recognised as having international impact and importance under the following criteria (Pereira, 2019; Kaur et al., 2019; Pereira & Van den Eynde, 2019):

1. Historic use for a period of at least 50 years;
2. Wide-ranging geographic application;
3. Utilisation in significant public or industrial projects;
4. Common recognition as a cultural icon, potentially including association with national identity or a significant individual contribution to architecture;
5. Ongoing availability for quarrying;
6. Potential benefits (cultural, scientific, architectural, environmental) arising from GHSR designation.

The HSR have designated the Cambrian age Welsh Slate which is quarried along strike in a large number of quarries located in Arfon as GHSR number 6 in 2018, and this is truly a stone of international importance having been widely exported in the 19th century and arguably 'roofed the World' during this century being exported across the British Isles and to Europe (and especially Germany) as well as to North America, New Zealand and Australia (Hughes et al., 2016). It is hoped that the Ordovician Slate of the Blaenau Ffestiniog area will achieve the same status in the near future. As of July 2021, Wales has four UNESCO World Heritage Sites in addition to the Castles and Town Walls of King Edward in Gwynedd; these are the Pontcysyllte Aqueduct in north-east Wales and the Blaenavon Industrial Landscape of Gwent in South Wales and the recently designated The Welsh Slate Landscape, encompassing all the slate quarry districts of North Wales (Welsh Government, 2020).

This series of volumes on Natural Stone and World Heritage will hopefully do much to increase both interest and knowledge of the importance of stone used in the constructions of buildings globally and throughout history and will be of use to architectural historians, archaeologists, builders, building conservators, geologists and the lay public. This specific study will hopefully increase the awareness of the GHSR stones as well as the less well-known building stones of NW Wales and their importance in the construction of the Castles of Edward I and other important buildings in this area. Therefore this book is concerned with both rock and stone, and these terms should be defined. I will use 'rock' to refer to geological materials *in situ* in the landscape and 'stone' to refer to material that has been removed from the natural environment and subsequently been manipulated by people.

A NOTE ON PLACE NAMES AND PRONUNCIATION

The place names in this book are the ones now used locally for the area. In the 19th and early 20th centuries, some Welsh place names were Anglicised, and the older literature in particular still uses these spellings; for example, Carnarvon and Conway have now reverted back to Caernarfon and Conwy and the latter name of these towns is used in the text. Similarly, the slate port midway between Bangor and Caernarfon was known as Port Dinorwic through the later 19th and 20th centuries. This name is still used but locally the town is more generally known as Y Felinheli, its original name before the slate barons took it over and again; this is the place name used herein. Beaumaris is referred to by its Latinised name, which is still widely used rather than the Welsh version Biwmaris. The ancient Kingdom of Gwynedd was subdivided into administrative regions known as Cantrefs, essentially representing one hundred settlements (*cant* meaning one hundred, *tref* meaning town). Although now largely lacking administrative power, these local regional names are still in use to this day. The area covered in this book lies within the cantrefs of Rhosyr (on Anglesey), Arfon (surrounding Caernarfon), Arllechwedd (Conwy and the Conwy Valley) and Dunoding (the region around the NE part of Cardigan Bay, which includes Harlech and the slate quarrying district of Blaenau Ffestiniog).

The Welsh language presents words and place names which at first glance appear impossible to pronounce to non-speakers. However, unlike English, the language is phonetic and letter sounds are mostly always the same wherever encountered. The Welsh alphabet includes double letter forms dd, ff and ll and the letters y and w are vowels as well as a, e, i, o, u. The most difficult letter for non-Welsh speakers to pronounce in Welsh is 'll'. The nearest equivalent in English is the 'tl' sound as in the word 'antler'. Therefore *llan* is pronounced something closer to 'tlan' rather than 'clan'. The letter 'u' is pronounced as an 'i'. 'Dd' is 'th' as in then (rather than th as in thistle). A single 'f' is pronounced as a 'v', whereas double 'ff' is pronounced as a single 'f' in English. 'W' is pronounced as the English letter 'u' and 'y' is pronounced either 'u' as in 'butter' or a short 'i' sound.

Acknowledgements

As a British geologist educated in the 1980s, I received an excellent grounding in Palaeozoic stratigraphy and tectonics of the British Isles and Western Europe. I am an alumna of the University of Birmingham, also the *alma mater* of Charles Lapworth, the geologist who defined the Ordovician period from his work in NW Wales and of University College London (UCL) where Edward Greenly studied geology in the 1880s. It has often been the case with geologists of my generation to rapidly migrate up the stratigraphic column into the sunny shores of the Cenozoic, leaving the murky, interminable greywackes of the Lower Palaeozoic far behind. This has certainly been my career path, but after two decades working on Neogene sediments in the eastern Mediterranean, it has been a delight to return to the geology of the western British Isles once more. I am therefore very grateful for the solid grounding in British stratigraphy that I received as an undergraduate at the University of Birmingham and of extending this knowledge west into Europe during my PhD programme at UCL. I am delighted at being able to return once again to the Palaeozoic to write this book, which for the first time has enabled me to contribute something to the geological story of the region in which I grew up.

Many thanks go to Dolores Pereira for proposing this series of books on Natural Stone and World Heritage and to the publishers Taylor & Francis Group for publishing this series. Many thanks go to my editor Alistair Bright, who has patiently endured me allowing this manuscript to expand in its scope over the last 2 years, during which much of my time has been eroded by a global pandemic which locked down libraries and restricted travel, limiting fieldwork and research.

However good one's skills as a geologist and a petrologist are, identifying building stones out of context from their geological outcrop is never an easy task. On showing many a geologist a rock sample, the first question that is asked will be 'where's it from?' Identifying building stones means one has to reverse-engineer one's thought processes to negate the requirement for this question. I am therefore grateful for beginning to learn this somewhat niche geological skill from Eric Robinson, my colleague at UCL for many years and a pioneer of the study of building stones. Returning to North Wales, I am very grateful for discussions with Andrew Haycock, Jana Horák and Evan Chapman of the National Museum of Wales and with members of the Welsh Stone Forum and particularly to David Roberts and Tim Palmer. Help with identifying stone and locating quarries has also been provided by Gareth Farr of the British Geological Survey, Dave Wallis of Geoscience Wales and Jayne McGrath of Midland Masonry.

Researching this book has inevitably involved visiting obscure quarries and monumental buildings as well as everything in between. As a non-driver, I am extremely grateful to Jane Siddall and Ruth Greenall for driving me to places that could not easily be reached using the local bus and train network.

Author

Ruth Siddall is a geologist, who applies analytical techniques from the field of Earth Sciences, and particularly petrology and mineralogy, to further the understanding of cultural material; pigments, construction and decorative stone, ceramics and cements in the built environment. Ruth has a BSc in Geology from the University of Birmingham and a PhD in tectonics and geochronology from University College London. She has over 25 years' experience of working on stone and building materials in a range of settings from Roman Corinth to Westminster Abbey. She is particularly interested in promoting geology and palaeontology through building materials and has written and led numerous building stone walks around London and elsewhere. For these activities she was awarded the Geologists' Association's Halstead Medal in 2019.

List of abbreviations and acronyms used in the text

BCE: Before Common Era; the equivalent of BC (before Christ)
BGS: British Geological Survey
Cadw: Welsh Government Historic Environment Service
CE: Common Era; the equivalent of AD (Anno Domini)
Ga: Unit abbreviation for a billion years (Giga-annum). When prefixed with a number, it means 'billions of years before present'.
GHSR: Global Heritage Stone Resource
HSR: Heritage Stone Resource
HSS: Heritage Stone Sub-commission
ICG: International Commission on Geoheritage
IUGS: International Union of Geological Sciences
ka: Unit abbreviation for a thousand years (kilo-annum). When prefixed with a number, it means 'thousands of years before present'.
Ma: Unit abbreviation for a million years (Mega-annum). When prefixed with a number, it means 'millions of years before present'.
NT: National Trust for Wales
NMW: National Museum of Wales, Cardiff
OS: Ordnance Survey
PLCM: Pennine Lower Coal Measures Group
RCAHMW: Royal Commission on Ancient and Historic Monuments for Wales
RIGS: Regionally important Geological Site
SSSI: Special Site of Scientific Interest
UNESCO: United Nations Educational, Scientific and Cultural Organization
WHS: World Heritage Site
WSF: Welsh Stone Forum

Chapter I

Introduction

The UNESCO World Heritage Site (WHS) that includes the Castles
and Town Walls of Caernarfon, Conwy, Harlech and Beaumaris
is a particularly large one, encompassing an 80 km section of the
North Wales coastline and its hinterland, which extends across the
coastal plains and up into the mountains of Snowdonia. It is spread
across three counties, Conwy, Anglesey and Gwynedd[1], and some
650 million years of Earth history. The Castles of Edward I and the
slate quarrying districts of Arfon and Blaenau-Ffestiniog (which
have also been nominated for WHS status) are located within a cultural landscape which has always had an intimate relationship with
rock and stone. For this book to provide a simplistic summary of
building materials over this large and geologically complex region
would be an almost impossible task. Therefore, to put the use of
stone in the Medieval Edwardian Castles into context requires an
understanding of what had come before and how the building of the
Castles and their immediate towns was influenced by this historic
landscape and how, in turn, these new settlements influenced the
regional development, right up until the present day.

North West (NW) Wales faces the Irish Sea. The towns of
Caernarfon and Conwy, some 35 km distant from each other, sit at
opposite ends of a coastal plain which extends between the Cambrian Mountains and the Menai Straits, much of this region is in

---

1 The administrative district of Gwynedd was named after the ancient Kingdom of Gwynedd, established following the fall of the Roman Empire. The
modern county designation was established in 1974 and until 1996 covered
the whole region under study. Gwynedd today comprises the old counties of
Caernarfonshire and Meirioneth.

DOI: 110.1201/9781003002444-1

the traditional administrative district (cantref) known as Arfon. Conwy lies in its own County, but the geology, west of the River Conwy, is contiguous with that of Arfon. In Welsh 'ar fon' means 'facing Anglesey', and this coastal strip is separated from the Island of Anglesey (Ynys Môn) by the NE-SW trending Menai Straits. The town and castle of Beaumaris are located in the NE of Anglesey on the Menai Straits, directly facing Arfon. Harlech Castle is an outlier, situated to the south of the Llyn Peninsula on the coast of Cardigan Bay, some 50 km south of Caernarfon. The Llyn Peninsula extends to the west separating Arfon from the Porthmadog and Harlech areas. The area described in the book covers the land west of the River Conwy to the central Llyn Peninsula, including the SW coast of Anglesey, parallel to the Menai Straits. It also includes the quarried regions of Snowdonia, Llanberis, Blaenau Ffestiniog and Dolwyddelan and the region around Harlech. The locations of the castles and towns discussed in the text are shown in Figure 1.1.

The regional geology of NW Wales comprises a sequence of Palaeozoic strata developed on an active continental margin and then, during the Carboniferous, a continental carbonate shelf. The geology of Anglesey is distinct from that of the mainland and is dominated by Proterozoic to Lower Palaeozoic igneous and metamorphic rocks and Carboniferous limestones and sandstones. Further south, Harlech belongs to a region of separate and distinct geology. The town of Harlech is located on the western side of the Harlech Dome, a broad anticlinal structure consisting of Cambrian to Ordovician deep shelf, clastic sediments of the central Welsh Basin.

This region of NW Wales includes some important building stones of both local and international significance. The Cambrian Slates of the Arfon region are designated a Global Heritage Stone (Hughes et al., 2016) and are certainly globally important building materials, and the Lower Carboniferous limestones and sandstones of Anglesey have regional importance as building stones and for the production of lime. In the last quarter of the 13th century and into the early 14th centuries, at the time of the building of Edward I's castles, these materials and other local, site-specific building stones were employed in construction of castles and churches, the main stone buildings of this period. However, many of the lithologies outcropping in NW Wales have been exploited for stone in some form or another, whether for building dry-stone walls (which are extensive in this region) or for the monumental-scale construction of the castles. With a long history of stone building dating back as far as

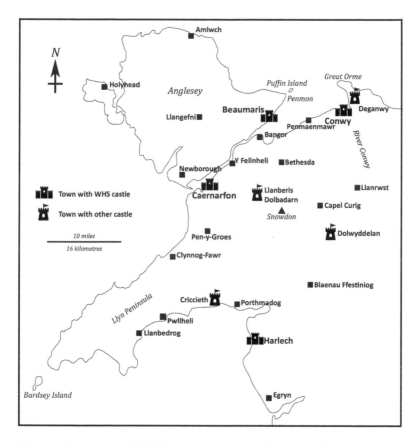

Figure 1.1 A map of NW Wales, showing the location of places named in the text.

the Neolithic, stone as precious resource was often recycled through many buildings, and even today, good stone is frequently salvaged from construction sites. The long coastline and good harbours have meant that movement of stone by sea routes has long been possible allowing for both export of local stone and import of stone from other areas of the British Isles. As a consequence, the identification of building stones is not so straightforward as one might think, even for a seasoned geologist with local knowledge. The main problem is that in provenancing stone, we are often searching for an absence in a landscape and not a presence. It is not unusual for stone resources to have been completely quarried out, leaving little evidence of a quarry, let alone remnants of bedrock for comparative purposes. This is the case with at least three lithologies considered important building stones in the Medieval period in NW Wales. Another local problem with the identification of building stones in the region is that the geological landscape of Snowdonia is dominated by a thick sequence of lower Palaeozoic metasediments and volcaniclastic rocks; different units can often be very difficult to distinguish once they are out of stratigraphic context.

Nevertheless, an understanding of stone and its properties has always been part of the culture of this region of Wales. The local landscape has long been a defensive one with a series of castles built in the region during the Norman and Medieval periods, and a millennium before that, the region was defended and garrisoned by a series of Roman forts. Prior to the Roman invasion, local communities in the Iron Age had constructed hill forts that could be defended in times of peril. Wooden forts and castles were constructed in both the Iron Age and the Norman period, and yet without stone, a castle is a vulnerable and an indefensible building. A wooden fort may be garrisoned but would be soon burned to the ground during a siege or indeed even a minor assault. Stone is required for truly defensive architecture.

The English King Edward I (1239–1307) was responsible for building the monumental stone castles at Caernarfon, Conwy, Harlech and Beaumaris. Edward spent most of his adult life at war. He succeeded his father, the relatively peaceable King Henry III, in 1272. At the time of his father's death, Edward was returning home from the Ninth Crusade. His wars against Wales began in 1277 when a short-lived campaign set out from the English border town of Chester. A full invasion of North Wales began in 1282 with the intention of total conquest and suppression of the local aristocracy. The building of a chain of castles was an important part of this military campaign, Edward fully understood both the defensive nature of castles and their ability to subdue and control local populations. Rhuddlan and Fflint Castles were the first to be built during the 1277 campaign, and the first to employ Edward's master castle builders were James of St George and Richard the Engineer. Both men were to go on to have a major part in the organisation of labour and materials as well as the architecture and construction of the four great castles of the Second Welsh War, the four new castles at Caernarfon, Conwy, Harlech and Beaumaris. Edward and his builders also modified and strengthened castles formerly in Welsh ownership, such as Criccieth and Dolwyddelan, and put garrisons in these structures. Rhuddlan and Flint Castles are outside the area of the current study. Located in the County of Clwyd, neither are they included in the UNESCO WHS.

Since the mid-19th century, NW Wales, with its rugged mountains and beautiful beaches together in close proximity, has been a popular tourist destination, and the castles of the Welsh Princes and Edward I have added to the attractions of the countryside and the castle towns. Subsequently, income for tourism has been of enormous economic importance to the region. The increase in the popularity in outdoor sports and activities, and particularly hillwalking, throughout the course of the 20th century has brought a huge influx of visitors to the region with an interest in the great outdoors and the natural history of the environment. Despite this, Snowdonia is one of the areas with the highest rainfall in the United Kingdom with an excess of 3,000 mm per year.[2] Visitors arriving with the intention to climb the mountains of Snowdonia are often driven to the lowland and coastal regions for alternative, wet-weather activities including the castles and other ancient and historic monuments in the care of Cadw and the National Trust for Wales. Visitor attractions focussed on geoheritage have recently also become popular destinations, from adventure activities and museums in the slate quarries at Penrhyn, Dinorwig and Blaenau Ffestiniog to the Bronze Age Copper mines on the Great Orme, near Llandudno. The island of Anglesey has been designated a UNESCO Geopark with an information hub at Port Amlwch. The reason for designating landscapes and monuments as WHS is two-fold. Primarily the intention is to protect them and ensure that they are appropriately managed and conserved but the designation also encourages tourism and brings visitors to these areas. NW Wales has an important place in the history of geology, and with this legacy and that of the minerals industry, it is beginning to embrace the concept of geotourism.

Tom Hose has defined 'geotourism' as 'the provision of interpretive and service facilities to enable tourists to acquire knowledge and understanding of the geology and geomorphology of a site (including its contribution to the development of the Earth sciences) beyond the level of mere aesthetic appreciation' (Hose, 1995, 2008).

---

2 Met Office UK, *https://www.metoffice.gov.uk/*.

This definition forwards the belief that landscapes, outcrops of rock and buildings of stone should be appreciated not for their awe inspiring qualities alone but that a means should also be provided for obtaining a deeper understanding of their geology and materiality, posing the following questions: What type of rock is this? How did it form? Why was it chosen to be quarried? At present, in NW Wales (as in many other locations), these questions are rarely answered in visitor centres where an emphasis is (not unreasonably) placed on the human story and aspects of engineering and transport infrastructure. For the slate quarries in particular, the engineering structures and particularly the influence of the slate industry on the development of the railways are perceived as the main focus of interest, with minimal description or discussion of the main purpose of a quarry, i.e. the stone which was won there. Far more has been written about the NW Welsh quarries because they were connected by the Ffestiniog Railway, and other lines, than there has been about the stone that was extracted from the quarries and its final destination. There are a number of slate museums within the region in both public and private hands, but as Price and Ronck (2019) point out, the geology is often seen as the 'background' with minimal (and not always accurate) geological or mineralogical information and displays are very much targeted towards industrial tourists rather than geotourists. A new geological attraction located in the heart of Snowdonia at the Ogwen Centre (Idwal Cottage) celebrates Charles Darwin's journey across Snowdonia in 1831. This wall is a beautiful piece of masonry, with coping of inlaid slabs of carefully selected, cut and often polished stones, depicting the landscape and geology that Darwin travelled across on his journey. Although a geological map is provided, there is not a key to the stones used, and they are therefore not identified or related to their place in the landscape, and they remain disconnected from the geological map (Snowdonia National Park, 2014). Once again, as Price and Ronck (2019) noted with reference to the slate tourism experience, an opportunity has been missed to appropriately present the geology.

There is potential for the geotourist visiting North Wales to be disappointed, and there are many, not least the legions of undergraduate geologists from British universities who learn to make geological maps in Snowdonia. This book hopes to further address this imbalance and put local stone at the forefront, facilitating its identification in buildings and tracing it back to the quarry from which it was extracted and telling the story of its formation as a geological material. This book, although it is hoped to be of interest to the geotourist, should also be a useful reference to architects, geologists and those involved in building conservation as well as readers with a broad interest in natural history and history.

The aim of this book is to equip the interested reader with the knowledge to recognise and interpret the use of stone in the buildings of NW Wales with primary focus on the WHS which encompasses the four castles of Caernarfon, Conwy, Harlech and Beaumaris. Chapter 2 provides an outline of the regional geology and stratigraphy with an emphasis on building stone resources. There has been a huge amount of literature written on the geology of NW Wales, albeit mostly within the academic corpus of writing. The area has been key to furthering the understanding of the evolution of active plate margins and oceanic plate settings during continental accretion. Important in the history of geology, the area of Wales in general has been significant stratigraphically, giving rise to the definition of the divisions of the Lower Palaeozoic; the Cambrian, Ordovician and Silurian, which are, as well as many of their stages, named from local toponyms. Similarly, much has been written on the quarries in the region, but this corpus of work has very much focussed on the history, engineering and archaeology of quarry sites. The authors of these works have largely ignored the geology of the quarried stone but also the destination and uses of the stone once it had left the quarry. Chapter 2 aims to summarise this knowledge and place quarries within their regional geological, historical and architectural contexts.

Chapter 3 provides a summary of the early history of stone building and building materials in NW Wales from the Neolithic period up until the death of Llywelyn the Great and the invasion of Edward 1 in 1277. The intention of this chapter is to provide context for the construction of Edward I's castles, which are described in terms of their building stones in Chapter 4. This is not the first text to illustrate the use of building stone in the castles of Edward I or indeed in North Wales. However, it is the first to look at these buildings within a long-term regional and temporal context of stone-building traditions in this region.

In many other texts, the castles are considered as single entities, and their materiality is described without the context of that of their surrounding towns. Looking at the use of stone in vernacular constructions has added much to the understanding of the building materials used in the castles. In a similar vein, the final Chapter 5 extends the use of building stone in NW Wales up until the 21st century describing the continuing traditions in building materials and how the introduction of new stones has influenced the regional character of the NW Welsh built environment.

Although it is not the purpose of this book to provide a detailed account of the history of NW Wales, it is hoped that enough detail is provided here to introduce readers to the sometimes turbulent history of this land and to be able to place the buildings and materials discussed within an historical timeframe. Those with a further interest in the regional architecture should turn to Pevsner's guide to Gwynedd (Haslam et al., 2009), which itself includes a brief introduction to the geology of building materials within the county. A series of detailed and authoritative essays on the phenomena of Medieval castles in North Wales and their place within culture and society has been edited by Williams and Kenyon (2010), and much of the work in this volume is cited within this text. This book is not intended to be a guidebook. For those wanting to explore further, the excellent new series of well-illustrated guidebooks produced by Cadw provide self-guided tours of the castles and other monuments. These are available for the four Edwardian Castles (Ashbee, 2015, 2017a,b; Taylor, 2015). Despite the well-exposed and interesting geology of North Wales, there are very few regional geological guidebooks available aimed at the layman. Roberts (1979)'s excellent and well-illustrated Geology of Snowdonia and Llyn is sadly out of print. Geologists' Association Guides are available for Anglesey (Bates & Davies, 1981) and the Llyn Peninsula (Cattermole & Romano, 1981) and the countrywide guidebook by Talbot and Cosgrove (2011).

Specifically relating to urban geology and the geodiversity of building stones, a number of town trails under the thematic title of 'Walking Through the Past' have been developed for towns in North Wales under the direction of Cynthia Burek at the University of Chester and the various Welsh RIGS (Regionally Important Geological Sites) Groups (see Burek, 2008). The latter organisations have also taken responsibility for the conservation of geological sites. In addition, the great majority of buildings, monuments and other stone-built structures mentioned in this text are also included on the website database London Pavement Geology created by the author and Dave Wallis in 2015. This resource was initially derived as an archive of London's building stones, but it has subsequently expanded to a UK-wide database, inviting submissions of building stones from localities throughout the United Kingdom (see Siddall, 2019). Over 350 locations are included for the region under study here (see London Pavement Geology http:// londonpavementgeology.co.uk/). On a national scale, the Welsh Stone Forum (Forwm Cerrig Cymru; see https://museum.wales/ curatorial/geology/welsh-stone-forum/.) has done much to research and promote interest and awareness of stone quarrying in Wales and the use of stone in Welsh buildings. Established in 2003 and based in the National Museum of Wales in Cardiff, this is a cross-disciplinary working group on building stone and includes membership from the fields of conservation, heritage, planning, geology, architecture and stone masonry who collaborate via regular meetings and field trips and publish reports in an annual newsletter. The National Museum of Wales in Cardiff also has a growing collection of Welsh buildings stones, which has been an invaluable resource in the re- search required for this book. Further information on the geology of Anglesey is available via UNESCO Geopark Geomôn's website at http://www.unesco.org/new/en/natural-sciences/environment/earth-sciences/unesco-global-geoparks/ list-of-unesco-global-geoparks/united-kingdom/geomon/. A large number of sites, quarries, castles and other buildings and monuments are described and referred to in this book. With the visitor in mind and for ease of reference, a gazetteer of sites and their locations (Ordnance Survey map grid reference) is included in this book.

The World Heritage Site of The Castles and Town Walls of King Edward in Gwynedd is listed by UNESCO at https://whc. unesco.org/en/list/374.

Chapter 2

Regional geology, building stones and quarries in North West Wales

2.1 INTRODUCTION

North West (NW) Wales has a long geological history, with an almost complete sequence of rocks from Precambrian (Neoproterozoic) to Upper Carboniferous age (Figure 2.1). Precambrian rocks outcrop on Anglesey, on the Llyn Peninsula and in Arfon, representing an oceanic plate setting and subduction zone margin environment. For most of the Lower Palaeozoic eras, the Welsh mainland was located on the northern margin of the Welsh Basin which accumulated a thick succession of turbidites and other clastic sediments. In the Ordovician, the region was once more an active continental margin during the Appalachian-Caledonian Orogeny with the development of a major igneous centre located in the Snowdonia region associated with extensive volcanism and the formation of caldera-derived pyroclastic rocks. Late Caledonian (Acadian) collision during the Silurian imparted a transpressive tectonic regime, which resulted in the formation of slate belts in Cambrian and Ordovician basin sediments. Upper Palaeozoic strata are dominated by regression in the Lower Carboniferous and the formation of carbonate platforms. Upper Carboniferous clastic sedimentary rocks occur in restricted basins in NW Wales, but these are far more extensively developed in the north east of the country, where they underlie the Permo-Triassic strata of the Cheshire Basin. Mesozoic and Cenozoic rocks are absent in outcrop in the region of NW Wales with the exception of a few, small Tertiary dolerite dykes and some Tertiary sediments exposed in the hanging wall of the Mochras Fault at Harlech. Regional uplift and erosion, influenced by the opening of the North Atlantic Ocean in

DOI: 10.1201/9781003002444-2

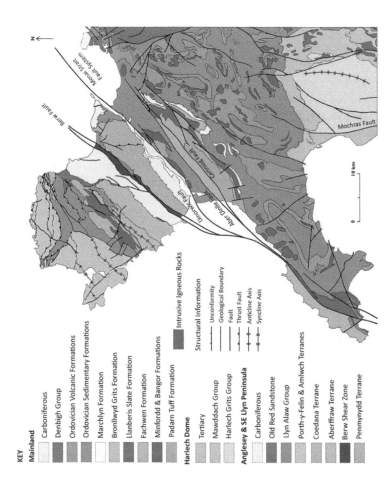

KEY

Mainland

Carboniferous

Denbigh Group

Ordovician Volcanic Formations

Ordovician Sedimentary Formations

Marchlyn Formation

Bronllwyd Grits Formation

Llanberis Slate Formation

Fachwen Formation

Minffordd & Bangor Formations

Padarn Tuff Formation

Harlech Dome

Tertiary

Mawddach Group

Harlech Grits Group

Anglesey & SE Llyn Peninsula

Carboniferous

Old Red Sandstone

Llyn Alaw Group

Porth-y-Felin & Amlwch Terranes

Coedana Terrane

Aberffraw Terrane

Berw Shear Zone

Penmynydd Terrane

Intrusive Igneous Rocks

Structural Information

Unconformity

Geological Boundary

Fault

Thrust Fault

Anticline Axis

Syncline Axis

Figure 2.1 The bedrock (solid) geology of NW Wales. (Modified from Schofield et al. (2020), Tectonic evolution of Anglesey and adjacent mainland North Wales, GSL Special Publications, 503 with permissions of the BGS © UKRI 2020.)

the Cenozoic, are largely responsible for creating the relatively high topography observed today. The mountainous region of Snowdonia was glaciated during the Pleistocene, and a well-developed sequence of deposits and landforms associated with the glaciation are spectacularly preserved in the region.

The regional geology is well exposed and relatively accessible. It is covered by the British Geological Survey's (BGS) 1:50,000 series geological maps, with bedrock and superficial deposit maps available (British Geological Survey, 1980, 1982, 1985a, 1989a, 1997, 2015). Much of the mountain region of Snowdonia and parts of Arfon and Conwy are additionally covered in the 1:25, 000 'Classical areas of British geology' series (see British Geological Survey, 1985b, 1986, 1989b, for regions relevant to this text). Field guides to the region include Bates and Davies (1981) Geologists' Association Guide to Anglesey, Roberts (1979)'s guide to Snowdonia and Llyn and Allen and Jackson's (1985b) guide to the Harlech Dome.

2.2 A HISTORY OF GEOLOGICAL RESEARCH

The region NW Wales is of global stratigraphic significance, due to it being the place of origin of the type sections of the Cambrian, Ordovician and Silurian Systems (Periods) of the geological timescale. These are named after *Cambria*, the Latin name for Wales and the *Ordovices*, and *Silures*, both Celtic-British tribes who occupied the region in the pre-Roman period. Field mapping and analyses of the regional geology began in the early 1830s conducted by Adam Sedgwick of the University of Cambridge, who was on occasion accompanied by Charles Darwin. However, Sedgwick mainly worked with Roderick Impey Murchison, who was at the time an amateur geologist (Howells et al., 1985). The Lower Palaeozoic successions of North and Mid-Wales are represented by a thick and somewhat monotonous sequence of poorly fossiliferous clastic sediments (greywackes) typical of continental margins, which in the early 19th century were undifferentiated as the 'Grauwacke [sic] Slates'. Sedgwick and Murchison (1835) defined the Cambrian and what was then understood to be the overlying Silurian System from their analyses of these clastic sedimentary strata and in so doing began one of the great controversies in the history of science which was to run for the next 50 years (see Thackray, 1976). The boundary between Sedgwick's Cambrian and Murchison's Silurian came under intense scrutiny, with geologists being unable to determine a clear paleontological change in the 'interminable greywacke' and finding that the same beds fell in both the Cambrian and Silurian Systems according to differing viewpoints. This led to an increasingly acrimonious dispute between once friends, Sedgwick and Murchison, each blaming the other for misinterpretation of the scant palaeontology, a fauna dominated by graptolites. The situation was resolved by University of Birmingham geologist Charles Lapworth, who stated the reality that 'the necessity for a tripartite grouping of the Lower Palaeozoic Rocks and Fossils, in partial accordance with this fact, has been very generally acknowledged' (Lapworth, 1879). Lapworth went on to point out that as Murchison's (by now increasingly disputed) account of the lower sections of the 'Silurian' was largely carried out in the land occupied by the *Ordovices*, which included 'Caernarvonshire' [sic], that this newly defined, middle division of the Lower Palaeozoic should be named the Ordovician System. Incidentally, the Ordovices's name also lingers on in local toponyms including the Iron-Age hillfort of Dinas Dinorwic in Arfon and the quarries (albeit in Cambrian Llanberis Slates) at Dinorwig. Nevertheless (and astonishingly), it was not until 1960 that Lapworth's designation of the Ordovician Period was formally ratified.

The British Geological Survey began mapping in North Wales in 1846 (see Allen & Jackson, 1985a; Howells & Smith, 1997), directly coincident with the discovery of gold in the Mawddach Group at Clogau near Dolgellau in Merionethshire in the same year. Mineralisation was an important economic driver for the progress of geological fieldwork, in addition to gold,[1] manganese and copper were being worked in the region as well as increasingly important building stone resources.

It was in the early 20th century that Edward Greenly began to contribute significantly to the understanding of the lithostratigraphy of the Island of Anglesey and the region of Arfon on the adjacent mainland. Having studied Geology at University College London, Greenly joined the British Geological Survey in 1889 and honed his field skills in the NW Highlands of Scotland. In 1895, he resigned from the Survey to pursue a personal project, the production of

---

1 Gold from Clogau Mine, the most lucrative gold source in the British Isles, has been used to make the rings for every British Royal wedding since that of Elizabeth Bowes-Lyon to George VI in 1923.

the first geological map of Anglesey. This was published in 1920 and remains in print to this day with only minor revisions (British Geological Survey, 1980). Greenly (1919)'s Anglesey memoir and his many publications on the geology of Arfon demonstrate his meticulous skills as an observational scientist and field geologist. Following his work on the Mona Complex of Anglesey, Greenly was also the first to coin the term 'mélange' to describe the chaotic deposits of the Gwna Group (Aberffraw Terrane), now interpreted as forming within the tectono-sedimentary environments of subduction margins. The complexities of Anglesey's geology did not come under further scrutiny until the later 20th century with advances in radiometric and geochemical studies. Revisions of Greenly's stratigraphy and structural analyses were undertaken by Barber and Max (1979), Horák et al. (1996) and Dallmeyer and Gibbons (1987). In recent years, a thorough reinterpretation of the geology of Anglesey has been undertaken using modern geochemical and radiometric techniques (see Asanuma et al., 2017; Schofield, 2020). Following the publication of his Anglesey magnum opus, Greenly turned his attention to the stratigraphy of Arfon and published a number of detailed accounts and geological maps of the mainland side of the Menai Straits (Greenly, 1928, 1938, 1944a,b, 1945, 1946).

Greenly was also one of the first geologists to take a serious interest in building stones and produced a survey of the castles at Caernarfon and Beaumaris in terms of their building materials (Greenly, 1932), and he also worked with archaeologists, including Sir Mortimer Wheeler and Wilfrid Hemp, in identifying the stones unearthed in excavations of the Prehistoric and Roman sites of Anglesey and North Wales. Further identifications and characterisation of building stones covering all the castles in North Wales, both pre- and post-conquest, were published by Neaverson (1947). Subsequently, there have been further revisions of the stones used in the castles constructed under Edward I's building campaign by Nichol (2005) and Lott (2010). These surveys have tended to look at the castles and their associated town walls in relative isolation from the surrounding settlements, and more generic studies of building materials both temporally and spatially distributed are few. Recently, Davies (2016, 2018a) has published summaries of the building stones in Welsh churches in the region.

The majority of rock types outcropping within the region have been used as building materials locally, regionally and, in the case of the Acadian slates, globally. The use of local stone in domestic architecture has done much to shape the regional character of North Wales. Nevertheless, we are reminded that this world heritage site covers a large geographical area and the stratigraphy and geology are far from being uniform across this region. The area can be broadly subdivided into three geological terranes: the Harlech Dome, Arfon and Anglesey which have somewhat distinct geological histories and stratigraphies. The scope of this book cannot provide a comprehensive stratigraphical account of the geology of region; however, it will present a summary of the strata outcropping in the regions of the four castles with emphasis on those materials which have had an impact on the built environment. Important building stones imported from outside the region are also included in this geological review.

On the mainland, the lower Palaeozoic sequence consisting of clastic sediments interspersed with periods of explosive volcanism has proven relatively straightforward to subdivide on a lithostratigraphic basis, despite dating of the 'interminable greywackes' proving difficult due to the scarcity of fossils throughout the region. This has resulted in almost constant revisions of formation and stage names over the last half century, making reading of the historical literature on the regional geology somewhat confusing. This text attempts to align itself with the stratigraphic divisions adopted at the current time by the British Geological Survey (2015). Nevertheless, where appropriate and within the context of building stones, alternative more commonly used names are occasionally used in this text.

2.3   THE PRECAMBRIAN ROCKS OF ANGLESEY

In North Wales, Precambrian lithologies are restricted to the Isle of Anglesey, and parts of the adjacent Arfon region and the southwest of the Llyn Peninsula are some of the oldest rocks exposed in the southern British Isles. The island is an amalgamation of a series of NE-SW trending geological terranes assembled along a south-east dipping subduction margin in the Neoproterozoic to early Cambrian periods. These were subsequently deformed during a Cambrian mountain building event which has been correlated with similar terrane histories in the Appalachians and is known as the Monian-Penobscottian Orogeny (Schofield et al., 2020). Recent geochemical and radiometric data have secured interpretation of the region as an oceanic plate and subduction zone sequence, representing the western margin of Avalonia during the Neo-proterozoic-Lower Palaeozoic closure of the Iapetus Ocean and subsequently accreted to the Lower Palaeozoic active continental margin of the North Wales mainland (Asanuma et al., 2017). The terranes of Anglesey have been recently revised by Schofield et al. (2020), and their terminology will be used in this text. From SE to NW, the terranes are named the Penmynydd Terrane, Berw Shear Zone, Aberffraw Terrane and Coedana Terrane. Outside the region of study are the Porth-y-Felin and Amlwch Terranes (previously known as the Monian Supergroup). The lithologies associated with these tectonic units are described below.

The Penmynydd Terrane (Schofield et al., 2020) extends in a tract occupying the SE portion of Anglesey, bounded by the Menai Straits Fault System to the SE and the Berw Fault to the NW. This unit is composed of metasediments at epidote to am-phibolite facies, resulting in spectacular micaceous greenschists as well as lenses of garnet-epidote-glaucophane metabasites. Two units have been identified: the Penmynydd Formation of blueschist-bearing metapelites and the overlying Pen-y-Parc For-mation which is dominated by greenschists. Garnet-glaucophane metabasites are exposed near Menai Bridge, and the Marquis of Anglesey's Column (constructed of Carboniferous limestone) is built on the most accessible outcrop of these rocks. Dates obtained indicating peak metamorphism for this terrane range between 575 and 550 Ma. The minimum age for the protolith is c. 585 Ma (Asanuma et al., 2017; Dallmeyer & Gibbons, 1987; Schofield et al., 2020).

North west of the Berw Shear Zone is the Aberffraw Terrane, which comprises the Bodorgan and Porth Trecastell Formations. The Porth Trecastell Formation (previously known as the Central Anglesey Shear Zone) is dominated by felsic, psammitic schists which have been strongly deformed by thrusting and folding. They include localised lenses of amphibolite and marbles. This forma-tion is overlain by the Bodorgan Formation, a mega-conglomer-ate: Greenly's original Gwna Mélange. The upper parts of this formation are dominated by the Llanddwyn Volcanic Member, which also affected by the Berw Fault is a mélange of pillow ba-salts, calcsilicates (rhodochrosite-bearing siliceous rocks) and red jaspers. A continuation of this unit is found on the northern coast of the southernmost Llyn Peninsula. A Cambrian-Ordovician Age has been assigned to this unit.

The oldest rocks on Anglesey are in the Coedana Terrane, which comprises the high-grade Coedana Complex Gneiss (666 Ma, Strachan et al., 2007; Horák, 1993), which is the host rock intrusive to the Coedana Granite (described under igneous rocks below). The Coedana Granite has been dated to 613 Ma.

The Proterozoic to Cambrian rocks of westernmost Anglesey are outside the area covered by this book, and although they are used locally for building (including for the construction of the impressive harbour at Port Amlwch), they are not important building stones brought to the eastern coast of Anglesey or to the mainland. Their stratigraphy has recently been revised by Schofield et al. (2020) as comprising the Porth-y-Felin and Amlwch Terranes. The Porth-y-Felin Terrane was previously known as the South Stack and New Harbour Groups, and these names will be retained here for the sake of simplicity. The intensely folded and thrusted South Stack Group is dominated by shelf sandstones and turbidites at low greenschist facies, whereas the superficially similar New Harbour Group additionally contains lenses of serpentinites, basalts and tuffs and has been interpreted as an olistostrome. Dates of 501 and 515 Ma have been determined by Asanuma et al. (2017) to constrain the deposition of the South Stack and New Harbour Groups, respectively. The serpentinites of the New Harbour Group have been worked as decorative stones. The Amlwch Terrane contains a similar series of lithologies to the Porth-y-Felin Terrane, but it is now understood to be separate. These units were previously lumped together as part of Gibbons and Ball (1991)'s Monian Supergroup.

All of Anglesey's Proterozoic lithologies are used locally and at a small scale for vernacular building and for structures including drystone walls. In one of the earliest surviving examples of monumental stone buildings in the area, the blueschists were used in the Neolithic (fourth Millennium BCE) for the construction of the passage grave of Bryn Celli Ddu, located near Llandaniel Fab in SE Anglesey. The Penmynydd Terrane has been quarried in the vicinity of Menai Bridge and Beaumaris and the greenschists of the Pen-y-Parc Formation of this unit are used in the construction of Beaumaris Castle and in local architecture (Figure 2.2). Blueschists and other schistose metasediments are mainly used in rough blocks for rubble masonry, though the schistosity lends itself to slab-shaped building stones. Blocks of jasper (metamorphosed siliceous rocks) and basalt, derived from the mélanges of the Aberffraw Terrane, have also found their way into rubble masonry construction.

Figure 2.2 (a) Greenschists from the Penmynydd Terrane used in early 19th century vernacular architecture in Beaumaris. (b) Blueschists from the Penmynydd Terrane used to construct the Neolithic burial chamber at Plas Newydd.

Serpentinites occur within the New Harbour Group which were, for a short time in the 19th century, worked as a decorative stone (Horák, 2005), and green varieties were marketed under the exotic sounding 'Verde de Mona'. These stones found a small market in ecclesiastical fittings and have been used in a number of British cathedrals. They were also used for chimneypieces; red varieties of so-called 'Mona Marble' were installed at Penrhyn Castle near Bangor (see Marsden, 2009; Horák, 2005).

2.4   THE NEOPROTEROZOIC TO CAMBRIAN ROCKS OF THE MAINLAND

The position of the Menai Straits separating Anglesey from the mainland is controlled by a series of northeast-southwest trending, steeply dipping faults, the Menai Strait Fault System. These structures represent a major boundary between the Monian Composite Terrane (Anglesey) and the Avalon Terrane (Mainland Wales), the western margin of the plate of Avalonia facing the Iapetus Ocean. Terranes were assembled during the closure of Iapetus during the Caledonian Orogeny. The Neoproterozoic to Cambrian was overall a time of marine transgression, with the Menai Strait Fault System having a strong control on basin formation and sedimentation. From west to east, the main structures in the Menai Strait Fault System are the Berw Fault (on Anglesey), the Dinorwic Fault, the Aber-Dinlle Fault and the Ceriniog Fault. In addition to separating the accreted terranes that comprise the Island of Anglesey and defining the trend of the Menai Straits, these structures have been active in controlling the tectono-sedimentary succession of the Arfon region and define a regional northeast-southwest strike. The metamorphic units described above on the Island of Anglesey only outcrop on the mainland in the westernmost Llyn Peninsula which lies outside the area of interest.

The sequence of volcaniclastic and clastic sedimentary rocks which straddle the Precambrian-Cambrian boundary is called the Arfon Group (the 'Arvonian' of Greenly, 1944a). In its type area of Arfon, these units are restricted to a region bounded by the Dinorwic Fault to the north west and the Ceriniog Fault Zone to the south east. The lower-most unit of the Arfon Group is the 1,000 m thick Padarn Tuff Formation. Its base is not exposed (Reedman et al., 1984; Howells et al., 1985). The Padarn Tuff outcrops between the Dinorwic and Aber Dinlle Faults in the area of Penrhosgarnedd, west of Bangor and also in its type area in a strip south west of Llyn Padarn, repeated by the Ceriniog Fault. It is a sequence of welded ash-flow tuffs (ignimbrite deposits) containing pumice, devitrified fiammé and quartz phenocrysts. Tucker and Pharaoh (1991) have assigned a Neoproterozoic age of 614±2 Ma (U-Pb zircon) to this unit. This stone is not quarried, but it was used for the construction of the Neolithic Bachwen Dolmen at Clynnog-Fawr and blocks of the tuff occur in glacial tills and subsequently in dry-stone walls and rubble masonry utilising boulders derived from glacial deposits.

The overlying Minffordd and Bangor Formations outcrop only in the vicinity of Bangor, NW of the Aber Dinlle Fault. There is also a small outlier on this unit on Anglesey at Beaumaris, the Baron Hill Beds of Reedman et al. (1984). These units form two prominent ridges which define the layout of the city of Bangor. The main city centre is located in the Adda Valley between the ridges, whilst the University Buildings are situated on the Upper Bangor – Bangor Mountain ridge to the north-west (Greenly, 1946). Approaching the city by rail, one passes through a tunnel through the south east ridge. Both Formations are composed of sandstones, conglomerates with clasts of volcanic rocks including blocks of Padarn Tuff as well as schists and basalts derived from the Aberffraw Terrane on Anglesey which grade up into acid tuffs and tuffites. These units have been used as building materials in Bangor but were never quarried on a large scale (Figure 2.3).

Figure 2.3 Bangor formation volcaniclastic rocks used for rubble masonry in the church of Our Lady and St James's (1866).

The Fachwen Formation outcrops between the Aber Dinlle and Ceriniog Faults, unconformably overlying the Padarn Tuffs. In many ways this unit is very similar to the Minffordd and Bangor Formations, but they are dominated by volcaniclastic sandstones with wedges of pebbly conglomerates and with notably fewer basalt clasts than those in Bangor. Howells et al. (1985) argue that the Fachwen Formation is probably the lateral equivalent of the Minffordd and Bangor Formations. Like the Padarn Tuff, the Fachwen pyroclastic rocks were not quarried but do occur in rubble masonry due to their incorporation into glacial deposits, especially in the vicinity of Bontnewydd and Glynllifon, to the south of Caernarfon. The Fachwen Formation passes up conformably into the Llanberis Slates, which are the most important building stones in the region.

2.4.1   Llanberis Slate Formation

The Lower Cambrian slates of Caernarvonshire belong to the Llanberis Group. In the Arfon region, the Llanberis Slate Formation overlies the Fachwen Formation sandstones. These are multi-coloured, metamorphosed mudstones, interbedded with a few subordinate, coarser-grained turbidite deposits. Colours vary from red through purples, blues and greys to rarer grey green and sage green slates, and the colour is imparted by variable amounts of iron and titanium oxides and chlorite. Red slates from the 'fengoch' (red vein) were particularly well-known from Dinorwig and Cilgwyn Quarries (Gwyn, 2015). Purple and red-purple varieties often have distinctive, pale-green reduction spots and veins of reduction along bedding laminations (Borradaile et al., 1991). The mudstones have been dated using the (extremely rarely occurring) trilobite *Pseudatops viola*, and Penrhyn Quarry at Bethesda hosts the type section for this unit. The lithology has been affected by a single phase of deformation resulting in homogeneous strain demonstrated by planar bedding and slaty cleavage. The ellipsoid reduction spots characteristic of some varieties of these slates have often been assumed to have originated as circular features, elongated during deformation. However, Nakamura and Borradaile (2001) believe that the reduction spots post-date strain and simply grew with a preference to the principal strain direction. Such slates have been marketed as 'Best Bangor Spotted and Striped' or *Goch Ysmotiog* (red spotted; Figure 2.4). This region of North Wales was deformed under a transpressional

Figure 2.4 Spotted Cambrian Welsh slate from the Llanberis slate formation. (a) Used for gravestones in Llanpeblig church-yard, Caernarfon and showing a large ellipsoid reduction spot. (b) Nantlle Slate used as cladding on a building on Caernarfon's Slate Quay. On the lower slabs, trails of reduction spots can be seen following original bedding. The slate has been split along the cleavage.

plate-tectonic regime in the closing phases of the Caledonian moun-tain building event, the Acadian Orogeny. This occurred in the Late Silurian – Early Devonian periods and was the result of a final, oblique collision of continental plates, the subsequent shear stress deforming the crust and forming the slaty cleavage. The slates have been extensively quarried in three main areas along strike, Nantlle, Llanberis (Dinorwig) and Bethesda and have been of huge economic importance to the region and have influenced the developments of settlements, both as 'quarry towns' occupied by workers and also the ports of Caernarfon, Port Dinorwic and Port Penrhyn at Bangor. Evidence of slate working for paving and roofing begins in the Roman Period, in the third century CE. Subsequently, the material was used routinely locally.

Caebraichycafn Quarry at Bethesda, once the biggest slate quarry in the World, and better known as Penrhyn, after its owners the Pennant family and its head, Lord Penrhyn, dominated slate working in the area of Bethesda. The Pennants monopolised slate quarrying in the lower Ogwen Valley. Work here probably started in the 16th century (Richards, 2007), but the quarries were bought out by Richard Pennant (1737–1808) in 1782. Pennant was the first to develop Welsh slate on an industrial scale. He came into possession of the Penrhyn estate on his marriage to Anna Susannah Warburton in the 1780s and opened the Penrhyn Quarry and developed Port Penrhyn (at Bangor). Pennant was a Liverpool man who had made his money in the slave trade and plantations in the West Indies. He was succeeded by a grand-nephew who took on his name, George Hay Dawkins Pennant (1763–1840), and who inherited both the quarries and plantations and was the builder of Penrhyn Castle. In turn, the quarry business was handed down to Dawkin-Pennant's son-in-law, Edward Gordon Douglas, Baron Penrhyn of Llandegai. In 1790, 400 men were working in the quarries, 100 years later 2,000 men were in the employ of the Pennants at Penrhyn Quarry alone (Jones, 1982). The Pennants were despised by their employees. They were everything the Welsh, Liberal, nonconformist quarrymen were not; English, Tories and Church of England. The huge Penrhyn Quarry was arranged on twenty-one galleries, each of which was served by a tram line. Teams of quarrymen would work their bargains from a set level, and slates were split and dressed on the levels too. Everything was transported to Port Penrhyn for shipment. The quarry is one of the few remaining in operation today, currently operated by Welsh Slate.

At Llanberis, early slate quarrying had also begun in the 16th century, and early workings, mainly relatively small pits, are scattered over the lower slopes of Elidir Fawr, west of the main quarry site. The earliest commercial mining and quarrying at Dinorwic began in 1787 driven by the landowner Thomas Assheton-Smith (1752–1828) of the Vaynol Estate and his son, also called Thomas (d. 1859), who opened up Vivian Quarry and then the sprawl of Dinorwig Quarry (an amalgamation of quarries called Wellington, Matilda, Victoria, Braich and Australia. The latter probably did supply slate to the cities of Melbourne but was so deep, it was probably named after what was perceived to be at its bottom). The Dinorwig quarries and the estate were inherited by Assheton-Smith's junior's niece on his death and subsequently by her sons, who took the Assheton-Smith name. Assheton-Smith senior built up the Dinorwic Quarry Company and built the railways down to his quays at Port Dinorwic on the Menai Straits. The first horse-drawn tramway to the purpose-built Port Dinorwic was built in 1824. Like Penrhyn, the industry was affected by strikes in the early 20th century and then the decline brought about by the First World War. Attempts to rejuvenate and modernise the slate industry were made, including opening the outlying Marchlyn Quarry, high above Deiniolen on Elidir Fawr in the 1930s. The Dinorwig Quarries continued to become less and less profitable and quarrying finally ceased in the region in July 1969. Subsequently, the area has been developed as a museum to Llanberis and Dinorwig's industrial heritage with a slate trail through the quarries and the slate railway (partially) renovated as a tourist attraction (The Padarn Railway). The old slate works at Gilfach Ddu are now the Welsh Slate Museum. The eastern part of the Dinorwig Slate quarries were redeveloped in the 1970s as Dinorwig Hydroelectric Power Station, fully commissioned in 1984 and capable of producing 1,728 MW of power. A network of tunnels and enormous machine halls were excavated into the mountain of Elidir Fawr, with twelve million tonnes of Llanberis Slate Formation removed. The rubble was used for landscaping the power station excavations and for related building projects. The remainder was tipped into the quarries and Llyn Peris; the rock flour has altered the colour of the water in comparison to that of the adjacent Llyn Padarn (Baines et al., 1983).

Across the valley to the south of the Dinorwig Quarry complex, a string of quarries mark the strike of the Llanberis Slate Formation across the landscape from Cwm-y-Glo to Nantlle. A series of relatively small-scale operations include the quarries at Moel Tryfan and Alexandra and numerous other smaller operations. These and the Nantlle enterprises were similarly relatively small and mainly owned by local or English businessmen who leased land for quarrying off the local landowners, the Wynn family of the Glynllifon Estate. Nevertheless, the Wynns maximised the number of leases with the intention of increasing their profit. However, this plan backfired and quarrying here never brought in as much money as the centralised quarrying systems operated in Bethesda and Llanberis. This was partly due to the piecemeal operation of many of the small quarries and the lack of a unified operation model. At Nantlle, Cilgwyn Quarry is generally regarded to be one of the oldest workings in North Wales, dating from at least the 13th century (Richards, 2007), and it may well have supplied slate for roofing and flooring at Segontium Roman fort and almost certainly supplied the slate for roofing Edward I's castles. It was the main quarry in the region producing slate until the early 19th century and the establishment and growth of Penrhyn Quarry at Bethesda, feeding the requirement for slate for the growing city of Dublin in Ireland.

Dorothea and Pen-yr-Orsedd are the biggest quarries in the Nantlle Valley. Dorothea opened in 1820 and was by far and away the most profitable quarry in the region. The land was originally owned by Richard Garnons (1774–1841), and the quarry was named after his wife. Dorothea was producing 500 tonnes of slate per year by 1848. The workings are impressive; Richards (2007) compares them to ancient Egypt, and indeed the great masonry piers erected to support inclines, steam engines to operate blondins[2] and a Cornish beam engine are reminiscent of temple pylons. Dorothea closed in 1968. Pen-yr-Orsedd opened in 1816 and was in operation until the 1970s. It too had blondins for lifting the stone from the pit, six in total which were initially powered by steam, but later electrified, they were in use until the quarry closed down. The Nantlle Railway opened in 1828 and was originally a horse-drawn tramway which later become steam-operated and ran down the valley to Caernarfon, from where the slate was shipped to Ireland and further afield (Richards, 2007).

Local landowners made fortunes from the slate, but working conditions for quarrymen and those with other craft specialisations, such as splitting and dressing, were poor, with very high mortality rates. The most useful sources on the social history of Welsh

---

2 Blondins were wire ropeways which crossed the open quarry pits and were used for transporting crates of slate. They were named after Charles Blondin, a famous French tightrope walker.

slate quarrymen are Jones (1982), Manning (2002) and Gwyn (2015), but Williams (1991) and (Richards, 2007) also contribute much on the working lives and practices of these men. The later 19th and early 20th centuries were marred by political unrest. The 1900–1903 strike and lock-out at Penrhyn, then the World's largest slate quarry, though unanticipated by both the workers and the owners, brought about the beginning of the decline of the Welsh slate quarries as global suppliers. Production dwindled to a minimum during the First World War and never really revived. The decline in supply of Welsh slate meant the increase in production at places such as Slate Valley in New York and Vermont States, USA (aided by the migration of Welsh Slate workers to this region; see Siddall, 2014), and the development of ceramic tiles as roofing material.

The Cambrian slates of Arfon were designated a Global Heritage Stone in 2016 and justifiably so, the slate trade which reached its peak in the late 19th century was truly global, with slate being exported from North Wales worldwide, from northern Europe (particularly Germany), to the Americas and Australia and New Zealand. It was at the time the most widely used roofing material, and Hughes et al. (2016) list a number of important buildings worldwide utilising the stone. It is, of course, the main roofing material used locally, but its use in North Wales was not limited to roofing slates. It was used extensively in all manner of applications. These include gravestones and memorials, for doorsteps, lintels, gateposts, window louvres, hearth stones, chimneypieces and billiard tables. West-facing exterior walls in the region are clad in slates, and it makes a very effective rain screen. Slates laid horizontally at foundation level also provided an efficient damp course. A style of fencing, unique to the region and locally called *crawiau*, incorporates upright slate slabs fixed in place by twisted wire. Within the quarry towns around Bethesda, a very localised folk art developed using slate to make stone books, fans and decorated, carved and inscribed fire surrounds (Caffell, 1983). Slate could also be 'marbled' with enamels so that it resembled more exotic stones. It must not be forgotten that slate writing boards and pencils were also used by every schoolchild for almost 200 years up until the early 20th century and the production of school slates and slate pencils, which were also exported globally, also proved itself to be a lucrative North Welsh industry (Gwyn, 2015; Richards, 2006; Williams, 1991).

The Llanberis Slate Formation is overlain by the mudstones, sandstones and conglomerates of the Bronllwyd and Marchlyn Formations.

These stones have not been quarried, though they are exposed in the stratigraphically (and topographically) higher levels of the Llanberis Slate Formation Quarries. Polished slabs of quartz conglomerates from the Bronllwyd Grits have been used to great decorative effect in the Darwin's Wall installation showing the potential of this stone.

2.5   CAMBRIAN AND ORDOVICIAN ROCKS IN THE HARLECH DOME

In the Harlech region, the strata are dominated by a series of Middle Cambrian greywacke sandstones, deposited as turbidites, the Harlech Grits Group and the Mawddach Groups. Indeed, it was in this area that the concept of turbidity currents and their distinct stratigraphic record was first understood and defined by the Dutch geologist Philip Kuenen (Kuenen, 1953). The Harlech Grits comprise a series of formations of siltstones, sandstones and conglomerates formed in the subsiding Welsh Basin and are described in detail by Allen and Jackson (1985b). The lowermost Llanbedr Formation is dominated by predominantly purple-coloured siltstones and mudstones which are overlain by the blue-green, gritty to conglomeratic facies of the Rhinog Grits Formation. A thin transitional buff sandstone, known as Egryn Stone, occurs locally in transitional boundaries of the Llanbedr Formation with the otherwise, greenish-grey Rhinog Grits, and this stone, though now entirely worked out, was a locally important building stone in the Medieval period (Palmer, 2007). Harlech Castle is built on a prominent ridge of Rhinog Grits sandstones, and this unit (including Egryn Stone) has provided the main construction materials for the Castle and other medieval and later buildings in the region (Figure 2.5). The overlying Hafotty Formation has a major manganese ore deposit at its base, hosted in a series of siltstones and sandstones. These beds pass up into the coarse-grained, quartz-rich greywacke sandstones of the Barmouth Formation before a return to mudstone/siltstone deposition in the Gamlan Formation.

The Mawddach Group represents a regressive sequence during a period of uplift and eventual emergence of the Cambrian Welsh Basin. Gold deposits are hosted in the black shales of the Clogau Group, formed in veins in association with a series of tonalite sills. The overlying Maentwrog and Ffestiniog Flags Formations are once more a series of siltstones and sandstones, and the uppermost

Figure 2.5 Rhinog formation sandstones used as buildings stones. (a) Typical, blue-green coloured Rhinog Grits used in the construction of Harlech Castle and (b) Egryn stone used as the dressings around a doorway at St Beuno's Church in Clynnog Fawr. Red Cheshire sandstone has been used to repair the door arch.

Dolgellau Formation is alternating siltstones and mudstones. The Ffestiniog Flags are the lateral equivalent of the Marchlyn Formation of Arfon. The Dolgellau Formation is composed of thinly laminated, black, pyritiferous mudstones lacking coarser-grained beds of sandstone marking a return to a period of deeper water deposition. The uppermost unit of the Mawddach Group, the Dol-cyn-afon Formation is assigned to the Lower Ordovician and is discussed below.

2.6   ORDOVICIAN STRATA

2.6.1   The Mawddach and Nant Ffrancon Groups

The Lower Ordovician represents a time period of initially high, but fluctuating sea levels followed by a major transgression in the early Upper Ordovician controlled by Gondwanan glaciation and locally, a period of intense volcanism (Howells, 2007). The British stage names for the Ordovician Series were originally defined and named after Welsh sections: the Tremadoc, Arenig, Llanvirn and Caradoc Stages. The uppermost Ashgill Stage was defined in northern England. These stages were originally defined based on graptolite biostratigraphy. The IUGS has more recently ratified divisions of the Ordovician based on global stratigraphy (see Bergström et al., 2009); nevertheless, it is appropriate to use the established British stratigraphic divisions in this area.

Lowest Ordovician rocks of Tremadoc age outcrop on the northern and eastern margins of the Harlech Dome. This is the Dol-cyn-afon Formation, the uppermost formation of the predominantly Cambrian Mawddach Group (Howells & Smith, 1997). This Formation is composed of grey to brown-coloured silty mudstones with interbedded sandstone horizons. It has a well-developed cleavage, which is most prominent in the mudstone facies. This unit was previously known as the Tremadoc Slates and represents an almost complete sequence of the Tremadocian stage with a biostratigraphy defined by trilobites, which though well preserved are typically deformed. The Dol-cyn-afon Formation is sub-divided into three members, the Lower Mudstone, Upper Sandstone and Upper Mudstone. The Upper Sandstone Member was previously known as the Porthmadog Flags, and these were quarried in and around the towns of Porthmadog and Tremadoc, as well as in Cwm Pennant (Dol Ifan Githin Quarry), and used for building stone (Figure 2.6). These are fine to medium-grained sandstones with cross-laminations, ripples and small-scale, channel cross-bedding. They represent a less stable environment than the black shales of the underlying Dolgellau Formation.

Figure 2.6 Dol-cyn-afon formation sandstones used for domestic buildings in Porthmadog.

Earliest Ordovician rocks in Arfon are the Nant Ffrancon Group, composed of siltstones and mudstones. These represent a major marine transgression following the deposition of the Mawddach Group. The base of the Nant Ffrancon Group is marked by a sandstone, the Allt Llwyd Formation which in Arfon unconformably overlies the Cambrian Arfon Group and the Twthill Granite intrusions and in the Porthmadog-Ffestiniog area, overlies the Dol-cyn-afon Formation (Howells & Smith, 1997). The Allt Llwyd Formation (the Graianog Sandstone Formation of Howells et al., 1985) is limited in outcrop and occurs as lenses at the base of the Nant Ffrancon Group (Figure 2.7). It has been assigned to the

Figure 2.7 Allt Llywd formation sandstones used in the early 19th century meat market in Caernarfon.

Arenig Series of the Lower Ordovician. In the area around Caernarfon, these sandstones are exposed lying on the eastern sides of the Twthill Granite.

They are also exposed in the town of Caernarfon, just south east of Twthill and more extensively just east of Crûg Farm near Bethel. This unit is also well exposed on Carnedd-y-Filiast in the less accessible mountainous region between Snowdon and the Nant Ffrancon Valley. They are medium to coarse-grained, well-bedded, weakly metamorphosed sandstones, varying in colour from pale grey to a dark grey green. Beds are around 15–25 cm thick and are well laminated. Bioturbation and films of organic matter are distinctive feature, and they contain the phosphatic oncolite *Bolapora undosa* (Howells et al., 1985). These sandstones are used in the second phase (early 14th century) of the building of Caernarfon Castle and in a few Victorian buildings in the town. These are Greenly's (1932 & 1944b) 'Basal Grits'. They were probably initially worked from the so-called 'Town-end' quarry on Twthill and were easily shaped into ashlar blocks, parallel to bedding (Greenly, 1932; Neaverson, 1947).

Above the Allt Llwyd Sandstone Formation, the majority of the Nant Ffrancon Group is composed of dark grey siltstones and silty mudstones which are variably pyritic or rich in phosphatic nodules. Towards the top of the formation, beds of pisolitic ironstones occur. Fossils are extremely sparse throughout the Formation and limited to a few graptolites. On the basis of these finds, the Nant Ffrancon Group as a whole has been placed in the Middle Ordovician, from the Arenig to the Llanvirn Series. The siltstones of the Nant Ffrancon Group are not important as a building stone in Arfon, although in the eastern Llyn Peninsula, locally silicified beds of green shales (see Roberts, 1979) are occasionally seen in rubble masonry and are used as cladding in some 19th century buildings around Llanaelhearn and Llithfaen. However, the story is very different for this unit on the eastern side of the Snowdon Massif. Here the Upper Ordovician Slate Belt has been as important to the economy of North Wales as the Cambrian slates of Arfon and has produced a series of very high-quality, although pyrite-bearing, grey to black slates (Figure 2.8). These slates have been quarried in a belt from Cwm Croesor, through Blaenau Ffestiniog to Penmachno. Here the Nant Ffrancon Group consists predominantly of blue-grey mudstones, siltstones and fine sandstones. In the large quarries and underground slate mines of Blaenau-Ffestiniog area, these are interbedded with pale-weathering, acidic ash-flow tuffs of the Rhiw Bach Volcanic

Figure 2.8 Black pyritiferous slate from the Nant Ffrancon formation quarried at Blaenau Ffestiniog and used for paving in Y Maes, Caernarfon in combination with purple, spotted Cambrian Slate from Penrhyn.

Formation. The slates are formed by a single penetrative cleavage present across all strata which is responsible for the development of the high-quality slates of this area. More often than not, cleavage is subvertical, striking NE-SW at Cwm Croesor. However, around Blaenau-Ffestiniog, cleavage dips at a low angle; 10°–35° and fans around in an antiformal structure to an E-W and then NW-SE strike, around the cleavage shadow related to the Tan-y-Grisiau Granite intrusion. At Blaenau-Ffestiniog, slate quarrying began on an industrial scale in the early 1800s and is still active today, although some quarries operate as tourist attractions too. At the height of operations, slate from the Blaenau-Ffestiniog region was contributing one-third of Wales's slate output (Richards, 2007). The main quarries are Oakeley, Diffwys, Llechwedd and Maenofferen Quarry complexes which lie to the north of the town. The Manod & Cwt-y-Bugail Quarries lie to the east in a fault bounded graben just E of the Manod dolerite intrusion. There are many other much smaller quarries scattered throughout the region. Richards (2007) records twenty-seven slate mines, quarries and pits. The biggest quarry was Oakley, though the scale of slate production in Blaenau-Ffestiniog was altogether on a smaller scale than at Penrhyn and Dinorwig. Oakley Quarry is around half the size of the enormous Penrhyn Quarry.

2.6.2   The slate quarries of Blaenau-Ffestiniog

As at Nantlle, the Blaenau-Ffestiniog quarries were in multiple ownership. Many operations were small. Landowners including Lord Newborough also operated quarry enterprises in the Ffestiniog area. However, the major quarry owners were the Oakeley and Greaves families. William Oakeley and his son, William Griffith Oakeley (1790–1835), began to develop land on the family's Rhiwbryfdir Estate in 1811. The Oakeley Family had been landed gentry in Shropshire and inherited their Ffestiniog estates through William senior's marriage to local heiress Margaret Griffiths. Oakeley first had discovered the economic potential of his land when he caught Lord Rothschild's agents prospecting for slate. Oakeley successfully sued Rothschild and began to establish their own quarry industry (RCAHMW, undated). They leased the land to a Liverpool stone contractor named Samuel Holland and his son, also called Samuel Holland (1803–1892), a man who was to become a significant player in the NW Welsh quarrying industry. Malchow (1992) provides a biography Samuel Holland juniors' life and business endeavours. The Hollands transformed what was now known as Oakeley Quarry into the largest and most profitable in the region and the largest underground slate mine in the World. Between 1820 and 1825, slate prices doubled, and by the time he was two in 1825, Holland junior had learned to speak Welsh and was turning a substantial profit for both himself and the Oakeleys. Holland always had an eye for business opportunities, and at the same time he bought a farm (which was sold on when it became a profitable concern) and was also instrumental in the foundations of the Ffestiniog Railway. He had also designed stone-cutting machinery. He carefully monitored the development of the quarrying industry and the demand for slate and indeed, other types of NW Welsh stone, but nevertheless had strong beliefs in liberalism and justice for Wales and its people – even if only to smooth out the complications of employing a great many local people. The Hollands sold part of their leasehold to Welsh Slate Company but continued to open up and own new workings on the Rhiwbryfdir Estate at Cesail Quarry (also known as Holland's Quarry; this has now been consumed by the greater Oakeley Quarry). Welsh Slate continued to manage Oakeley Quarry well, and the Oakeley family continued to take royalties on all profits. When the leases came up for renewal in 1878, both for Welsh Slate and for Holland, the then landowner William Edward Oakeley (1828–1912) refused to renew it and himself took control of the slate business, consolidating operations into the Oakeley Quarry company which was to continue working the quarries alongside Welsh Slate (albeit through some very trying times, which included major rock falls, strikes and the 20th century wars) up until 1969 (Richards, 2007). However, Oakeley was never to be as profitable as it had been under Holland or Welsh Slate. In 1880, Samuel Holland junior went on to stand for Parliament and became the Liberal MP for Meirionethshire. Holland lived at Plas-yn-Penrhyn in Penrhyndeudraeth (Malchow, 1992).

John Whitehead Greaves (1807–1880) was from very different stock than the Oakeleys. He came from a Liberal, English Quaker family of bankers. Greaves did not go into the family business but came to Caernarfon in 1830 and developed an interest in the slate industry and began to prospect the land for potential slate bonanzas. He opened Llechwedd Quarry in 1846 and exhibited his slate at the Great Exhibition of 1851, where he received an order to roof Kensington Palace. The slate business was inherited by J. W. Greaves's son, John Ernest and by this time in addition to Llechwedd included Diffwys and Maenofferen quarries. J. W. Greaves Ltd. still exists today, operating Llechwedd as a tourist attraction alongside small-scale slate quarrying (Richards, 2007). Cwt-y-Bugail is currently operated by the new Welsh Slate company.

Ffestiniog Slate was transported to Porthmadog first by road and then boat along the River Dwyryd and later, from 1836 by the Ffestiniog Railway. The slate was often known at the time as Porthmadog Slate. Although the work was as a demanding and dangerous as at Penrhyn and Dinorwig, the many quarries, many owners and a continual demand for workers meant that quarrymen could easily leave one quarry and find employment in another, and so there was generally less dissatisfaction within the workforce. This meant that trade unions developed to a far lesser degree than in the Arfon quarries. Industrial action did occur but nowhere near on the scale of the Penrhyn lock-out of the early 1900s.

Slate from the Blaenau-Ffestiniog area was brought down to the port of Porthmadog via the Ffestiniog Railway, which had been established as early as 1833 as a horse- and gravity-driven line.

2.6.3   The Middle Ordovician Strata of Anglesey

On Anglesey, Arenig-Llanvirn age transgressive sandstones, gritstones and conglomerates outcrop in the north-west of the island, outcropping in Y-shaped area (see Figure 2.1) between the Neoproterozoic tectonic terranes and in several outliers around Beaumaris and elsewhere in the island. At the time of writing, these units are undergoing reclassification which will inevitably lead to confusion in terminology, and so they are broadly described here using the revised stratigraphic group nomenclature of Schofield et al. (2020). Ordovician strata have been renamed as forming the Cemaes, Llyn Alaw and Porth Wen Groups. These are not correlated with the Ordovician strata of the mainland and although having suffered some deformation, they are a uniformly low metamorphic grade. Lower Ordovician Cemaes Group sediments are predominantly volcanic and volcaniclastic in origin, with a sequence of greenish-coloured sandstones and tuffs overlain by a conglomerate (the Port Swtan Formation). Middle Ordovician sandstones and conglomerates dominate the Llyn Alaw and Porth Wen Groups. Conglomeratic sandstones contain prominent, angular clasts of Mona Complex lithologies, including schists and jaspers in a brown or blue-grey matrix (see Neuman & Bates, 1978). Late Ordovician strata outcrop around Amlwch and a succession of volcanic, rhyolitic composition rocks host the copper ores of Parys Mountain.

The Ordovician strata of Anglesey are not important as building stones in the area covered by this book or indeed more widely in Anglesey. They are largely covered by a thick blanket of Pleistocene to Holocene superficial deposits and therefore outcrop is best observed in coastal sections. The largest outcrops of these rocks in the region under study are the grey, mudstones and sandstones of the Baron Hill Formation (Llyn Alaw Group) which outcrop in the vicinity of Beaumaris; however, stone from this formation is not used in the town as a building material. Ordovician sandstones are used locally as stones for dry-stone walls and conglomeratic facies of these units were used in the Neolithic for the construction of Burial Chambers at Ty Newydd and Presaddfed, in the latter case these were presumably large blocks from glacial deposits. Green sandstones from the Cemaes Group have also been identified by Horák (2013) as materials used for inscribed stones on Anglesey in the Early Christian Period (5th–7th century). Copper has been worked at Parys Mountain since the Bronze Age, and the mine complex was one of the world's major producers of copper in the early 19th century.

2.6.4   Caradoc volcanic rocks and slates

During the Middle Ordovician, what is now North Wales became an active continental margin on the Avalonian continent during the closure of the Iapetus Ocean and the Caledonian Orogeny. Tuff layers, indicating the onset of explosive volcanism, start to appear within the Nant Ffrancon Group; however, in the Late Ordovician Caradoc Series, the development of major acidic eruptive centres occurs in what is now the upland area of Snowdonia. The Llewelyn Volcanic Group was probably erupted from an igneous centre in the region of the Glyder mountains. These units have the most use as building stones in the area of Conwy and the Upper Conwy Valley. The Conwy Volcanic and Capel Curig Formations are both composed of a series of ignimbrite deposits; acidic ash-flow tuffs, weathering buff to white and showing varying degrees of welding and the presence of devitrified fiammé. The Conwy Rhyolites outcrop trending NNE-SSW from Foel Fras to Conwy, attaining a maximum thickness of ~1,000 m. On Conwy Mountain, the unit has been metasomatised, with sericite replacing feldspars. The overlying Capel Curig Volcanic Formation similarly outcrops over a large area from the Capel Curig area to Conwy; it represents a major ash-flow tuff, up to 400 m thick. The Capel Curig and Conwy Formations lie in close proximity in the towns of Conwy and Deganwy and are a distinctive lithology used in the Medieval Town walls of Conwy and for the construction of the early Medieval Castle at Deganwy, which itself sits on a prominent outcrop of Capel Curig Formation ignimbrites (Figure 2.9). A quarry is located in the rhyolitic rocks of the Conwy Volcanic Formation on the north side of Conwy Mountain (Mynydd-y-Dref).

The Llewelyn Volcanic Formation is overlain by the extensive Snowdon Volcanic Group which dominates the uplands of Snowdonia. The two units are separated by the Cwm Eigiau Formation, a sequence of mudstones and siltstones, with occasional tuff beds, which have subsequently developed a slaty cleavage. Several quarries work these stones in the Llugwy Valley between Betws-y-Coed and Capel Curig and primarily at Hafod Las Quarry which supplied slab building stones locally and particularly for the Victorian development of Betws-y-Coed (Gwyn, 2015; Richards, 2007). At Dolwyddelan in the Lledr Valley, the slate is very dark grey and black, but of poorer quality to that from the Blaenau-Ffestiniog District. Dolwyddelan Slate was quarried from narrow, steeply dipping veins, within the ENE-WSW trending beds of the Nod Glas Formation, exposed in the core of the Dolwyddelan Syncline; '…here managers and men were chasing narrow and faulted veins of indifferent slate' (Gwyn & Davidson, 1995). Industrial-scale quarrying began here in the 1820s, though this stone had been

Figure 2.9 Conwy rhyolite used in the construction of the Town Walls of Conwy. The felsic rhyolite stained with iron oxides.

worked in the early 13th century for the construction of Dolwyddelan Castle, along with the overlying rhyolitic tuffs of the Snowdon Volcanic Group.

Resurgence of volcanic activity in the Upper Caradoc occurred through the development of the Snowdon Caldera and the eruption of the Pitts Head and Rhyolitic Tuff Formations; major ignimbrite sheet flows dominate the geology of the mountainous hinterland of Snowdonia. These units were seldom quarried, though field stones were used for walling and sheep folds and their distinctive eutaxitic texture is striking in buildings constructed from glacial cobbles. The geology of the Gwydir Forest area of the west bank of the Conwy Valley is dominated by the locally named Crafnant and Dolgarrog Volcanic Formations also of the upper Caradoc Snowdon Volcanic Group. Despite the remote locations of this area, some thirty, small, quarries have been located which work the cleaved mudstones interbedded with the volcanic rocks including Clogwyn-y-Fwch where very iron-rich, bright orange-stained slates were mined and also at the remote mountainous locations of Cedryn and Cwm Eigiau. Inevitably pyroclastic rocks, tuffs and tuffites occur frequently in the building stones in the Conwy Valley. It should also be noted that this region was also intensively mined for copper, lead and zinc in the 18th and 19th centuries.

Extensive tramways and inclines were built to transport slates from various mines and quarries in the Gwydir Forest region to the Conwy Valley and then to the wharfs at Trefriw, located at the limit of navigation of the Conwy River, where stone could be shipped to Conwy. Much has been written about the infrastructure supporting the transport of these stones (see Richards, 2007; Gwyn, 2015). However, the Conwy Valley slates were clearly inferior roofing slates compared with those derived from the Cambrian Strata of Arfon or from Blaenau Ffestiniog, and it is not clear where this export market was outside the construction of towns and villages in the Conwy Valley.

2.6.5   Conwy Castle Grits

In the vicinity of Conwy, Upper Ordovician, Ashgill Series clastic sedimentary rocks crop out in the form of the Conwy Mudstone Formation; a sequence representing basin shallowing. These are fine-grained siltstones and mudstones and are not locally important as building stones in the main part. However, a prominent sandstone to conglomerate facies, the Conwy Castle Grits Member is part of this unit. The Conwy Mudstone and Castle grits outcrop in a series of anticlines and synclines south and west of the River Conwy. The Conwy Castle Grits are brown to dark grey medium-grained, calcarenitic sandstones, which distinctively contain beds with rounded pebbles and rip-up clasts, up to around 8 cm in length, either isolated or in strings (Howells, 2007). The stones were quarried from the castle rock itself, as well as in the area of the village of Gyffin, 200 m south west of the main town (Neaverson, 1947). This stone is the main building stone used in Conwy, including in the Castle and town walls and also in the surrounding villages in the upper Conwy Valley.

2.7   SILURIAN AND DEVONIAN

The end of the Ordovician brought about a global rise in sea level which was dominant throughout the Silurian. However, tectonic activity associated with the closing stages of the Caledonian Orogeny brought about regional sea-level fluctuations and eventually emergence of the landmass in the Devonian. This period, the closing Acadian phase of the Caledonian Orogeny, was also important for the transpressive tectonic regime which imparted the regional slatey cleavage in the Lower Palaeozoic clastic sediments. Silurian strata are poorly represented in the region under study herein and have limited use as building stones west of the Conwy Valley. However, they outcrop on the eastern margins of the Harlech Dome, on Anglesey and in the Conwy Valley. Lower Silurian black shales outcrop in the vicinity of Conwy and near Parys Mountain on Anglesey. In the Conwy Valley the Wenlock-age Denbigh Grits are a sequence of clastic sediments deposited as turbidity currents in a basin possibly controlled by the NNW-SSE trending Conwy Valley Fault (see Howells, 2007). These rocks are a sequence of sandstones, siltstones and mudstones, the latter with a distinctive striped texture. Sandstones range from fine- to coarse-grained and conglomeratic facies. A number of small quarries have been located between Tal-y-Cafn and Maenan on the eastern bank of the River Conwy, and these were probably initially exploited for building the 13th century Maenan Abbey and the 18th and 19th century Maenan Abbey manor houses. The Denbigh Grits pass up into the Nantglyn Flags Formation. Richards (2007)'s diligent survey of slate quarries records fourteen quarries located in the lowermost units of the Nantglyn Flags which outcrop above Llanrwst and Bodnant on the east side of the Conwy Valley, most of these workings date (where such information is available) from the 18th and 19th centuries, and many are small pits with limited exploitation. These stones have been used for local building in the Conwy Valley. The Nantglyn Flags are actively quarried, outside the region of interest of this book, on the Horseshoe Pass (Denbighshire) and are marketed as Berwyn Slate, though they have limited use as roofing slates, and as in the last 500 years, the material is mainly used for flooring, cladding and countertops (Figure 2.10).

Devonian strata, the 'Old Red Sandstone', are restricted to a narrow outcrop on Anglesey and are represented by continental red beds of red-purple sandstones and mudstones exposed between Dulas and Lligwy Bay, underlying the Carboniferous sandstones and limestones of the Clwyd Group in the vicinity of Moelfre (Greenly, 1919). These stones have not been observed in the construction materials of local churches or other buildings (Davies, 2016).

Figure 2.10 Conwy Castle Grits sandstone member used in the construction of Conwy Castle. A string of mudstone pebbles can be seen in this sample.

2.8 THE CARBONIFEROUS ROCKS OF ANGLESEY AND THE MAINLAND

The Lower Carboniferous (Mississippian) rocks of NW Wales are probably the most important buildings stones in the region and are classified as the Clwyd Limestone Group. They fall into the upper Dinantian Series (Visean Stage) which in the British Isles is sub-divided into the Arundian, Holkerian, Asbian and Brigantian Sub-stages. These strata of late Holkerian to Brigantian age have been quarried on an industrial scale both for dimension stone and as a source of lime in the Penmon and Moelfre quarry districts located in the north-west of the island of Anglesey and also on the mainland at Vaynol and Llandudno (Great Orme) and extensively at numerous sites in Clwyd. Stratigraphically, they comprise a series of basal, often conglomeratic sandstones and four overlying limestone formations. These basal conglomerates are known by a series of local names, and they are very variable on a local scale. These units are poorly known geologically but are extremely important as building stones throughout NW Wales and beyond. As a building material, they are generically referred to as the Basement Beds in this text, though local facies variations and quarry sites are described where appropriate. Up until the 18th century, these sandstones were far more important as a building stone than the limestones which were mainly used for the production of lime. By the 18th and 19th centuries, limestone and lime were the main products of the Anglesey quarries, and these stones gained more importance as building stones, first in the town of Beaumaris, and later they were also used in major engineering projects, including the construction of docks and harbours across the region and for the bridges crossing the Menai Straits and the Conwy Estuary. The lowermost of the limestones, the Leete Limestone, is only exposed at Penmon in north eastern Anglesey and the nearby quarry at Tandinas, just north-west of the Penmon Quarry district. The overlying Loggerheads Limestone also contains numerous sandstone horizons which have been exploited for building stone, and it is in itself a high-quality construction limestone. The Cefn Mawr Formation is the highest in the sequence exposed at Penmon. The uppermost formation in the Clwyd Group, the Red Wharf Limestone Formation, is exposed in the Moelfre quarry district and on the Vaynol Estate near Bangor on the mainland. In this latter area, stone was only quarried on a small scale for local use. Given their importance as building materials, these geological units and the stones are described in more detail below.

2.8.1   The Basement Beds

The term Basement Beds generically and informally refer to a number of outcrops of pebbly sandstones and gritstones which are easily overlooked in terms of the regional geology of Arfon and Anglesey but are of huge importance as local building materials. It is understood that the term 'Basement Beds' is largely obsolete in the modern geological literature, and these units are now named locally as members of the Clwyd Limestone Group. Some of them indeed do form the base of the limestone sequence, but they are also interbedded with the limestones, especially in their lower parts. However, within the context of a study of building stones, it is useful to group the various outcrops and quarry localities of these sandstones, gritstones and conglomerates as a 'stone name', and therefore they will generically be referred to as the Basement Beds in this work. Generally assigned to the Visean stage and Holkerian-Asbian sub-stages of the Lower

Carboniferous, these sandstones partially underlie the Clwyd Group Limestone (hence, the original designation as Basement Beds) but are also interbedded within the Loggerheads Limestone (Davies, 1982; Waters et al., 2007). Lithologies range from uniform sandstones to coarse gritstones, conglomerates and breccias with clasts predominantly of vein quartz, but clasts of schists and red jaspers are also common, indicating a provenance from the Neoproterozoic and early Cambrian terranes of Anglesey. Outcrops on the mainland shore of the Menai Straits are known to contain boulders of greenschist derived from the Penmynydd Terrane of Anglesey (Greenly, 1928). The sandstones vary in colour from buff to yellow, red, purple and green, relating to various oxidation states of iron oxide minerals disseminated throughout these rocks. Bedding and cross-bedding structures are often present and well developed. In outcrop, these sandstones are limited to cuvettes each with their own character and are now named locally as distinct stratigraphic units, formalised as Members in the Loggerheads Limestone Formation (Figure 2.11). However, as ever, stratigraphers can be designated 'lumpers' or 'splitters' in their division of geological strata. Davies (2003) lists the following localised formation names of these strata on Anglesey; on the north east coast these are the Benllech Sandstone, Helaeth Sandstone, Lligwy Sandstone and Lligwy Bay Conglomerate. The Pencraig Sandstone outcrops near Llangefni. At Penmon these are known as the Fedw and Parc Sandstones. On the Anglesey coast of the Menai Straits, outcrops are called the Moel-y-don Sandstone, Carnedd Sandstone, Edwen Sandstone and Fanogle (or Fanogl) Sandstone. Basement Beds also outcrop on the mainland Menai Straits shore on the Vaynol Estate and are known here as the Loam-Breccia Formation where Greenly (1928) recorded them as having a distinctive lavender-purple colour and containing large clasts of Penmynydd Terrane greenschist facies rocks. This unit was later designated the Bridges Formation, as geographically, it is mainly confined between the Menai and Britannia Bridges. More recently, Waters et al. (2007) have assigned the Anglesey formations to a single Lligwy Bay Formation and the mainland breccias into the Menai Straits Formation (which includes outcrops on the Anglesey shore of the Menai Straits). It is true to say that a greater variety of stone in terms of colour and sedimentary facies has been observed in buildings than in outcrop, suggesting that some facies have been entirely worked out.

Figure 2.11 Sandstones, gritstones and conglomerates from the Basement Beds group of Lower Carboniferous sandstones.
These are the most important building stones in Anglesey and Arfon from the Roman period through to early
19th century. There is considerable variation within these
sandstones, and many can be located to quarry source.
(a) Typical quartz conglomerates, with iron stained matrix used in the wall as Doc Fictoria, Caernarfon. (b)
Malltraeth facies of fine to medium-grained, dolomite cemented freestones at Plas Bowman, Caernarfon. (c) Red
sandstones from Moel-y-Don. This boulder at St Nidan's
Church shows coarse and fine-grained sandstones, ranging in colour from pink to liver-red. (d) Cross-bedded,
green gritstones used as window dressings in Caernarfon
Castle. (e) Lavender-coloured sandstones conglomerates
and a quartzite used in Caernarfon's Town walls. The lavender coloured sandstones come from the mainland shore
of the Menai Straits at Vaynol (Bridges Formation). (f) Red
jasper in quartz gritstone used at Beaumaris Castle.

Despite their frequently rubbly and unwieldy appearance, the basement beds gritstones and conglomerates have been widely used as building stones in Anglesey and Arfon from the Neolithic onwards. Perhaps unsurprisingly they were used for the construction of millstones, but the more uniform varieties were suitable for use as freestones employed in sculpture from the Roman Period onwards. Millstones were quarried at Trwyn Ddu and Fedw Fawr near Penmon in the 17th century (Thomas, 2014). The sandstones were quarried alongside the limestones at Penmon and in the Moelfre and Benllech regions of Anglesey (see below) but were also worked at all outcrops for local use. One may argue that in the Medieval period, that the limestones were a by-product of the sandstone quarrying industry, the latter primarily being used for lime rather than masonry. Many facies of the Basement Beds are rubbly, coarse-grained, poorly sorted conglomerates and breccias, although quartz and vein quartz clasts are predominant, the occurrence of bright red Mona Complex jasper is very distinctive. These varieties are less easy to locate to a specific quarry but are part of the Lligwy Formation. Very distinct facies occur at Moel-y-Don on the Anglesey coast opposite Port Dinorwic, where they vary in colour from a dark red to pink colour as well as mottled cream and red varieties. These red sandstones were quarried on land belonging to the Augustinian Monasteries. The stone has been almost entirely quarried out, leaving a low-lying, tongue-shaped promontory with scant outcrop still *in situ* (see Shipton, 2009). At Malltraeth on Anglesey, a buff-coloured, medium-grained, well-sorted dolomitic sandstone outcrops and was quarried there. It also outcrops between Llangefni and Llanbedr Goch and at Foel on the east coast of the island (David Roberts and WSF, *Personal Communication*; Davies, 2018a). This particular facies of stone, despite containing the occasional sub-rounded quartz pebbles, up to 1 cm long, had properties approaching that of a freestone and was consequently used both for sculptural and dimension stone from the Roman period onwards. This Malltraeth Sandstone was a prized stone that was widely used across the region and beyond. Also in the Llanbedr Goch and Creigiau areas, white and grey, coarse-grained, quartz arenites (quartzites) outcrop. Creigiau facies quartzites are occasionally seen in buildings, including at Beaumaris Castle but are less commonly used than other facies of the Basement Beds.

Basement Beds grits are used extensively in the walls of Roman Fort of Segontium, at Caernarfon and Beaumaris castles and town walls and in churches across the region in question of the same period, including Bangor cathedral. They are also widely used in domestic architecture across Arfon and Anglesey and exported throughout North and Central Wales. According to Davies (2016), they were exported and used as building materials at the foremost Cistercian abbey complex at Strata Florida in Ceredigion, some 160 km to the south of Caernarfon. The Moel-y-don and Malltraeth facies of the basement beds have been identified, particularly at Strata Florida. These were probably acknowledged as the finest building stones locally available and both had properties and potential as freestones.

2.8.2   Clwyd Limestone Group

The North Welsh continental shelf developed over the Visean (Middle Pennsylvanian) depositing the series of transgressive-regressive cycles of the Clwyd Limestone Group. These limestone units lie unconformably on the Basement Beds or directly on Silurian or (in Anglesey) Devonian strata. These limestones outcrop in Clwyd in NE Wales, on the Great Orme and in NE Anglesey at Moelfre and Penmon as well as small outcrops along the Menai Straits. The Clwyd Group of Anglesey has been described in detail by Davies (1982) and has been regionally summarised by Waters et al. (2007). The Clwyd Limestone Group is subdivided into four major formations, which broadly demonstrate a transgressive sequence. The lower units of Leete Limestone Formation (Holkerian to Asbian age) are grey, often brown and rusty weathering, argillaceous, shaley limestones with numerous fossils of the brachiopod *Daviesiella llangollensis*. These were deposited in a peritidal lagoon environment. The upper member of this Formation consists of distinctive, white-weathering, porcellanous limestones with variable distribution of vuggy, bird's-eye texture. This particular facies was worked from Tandinas and Flagstaff Quarries at Penmon in Anglesey and also in the vicinity of Benllech in the Moelfre district and is known as Tandinas Stone. Flagstaff Quarry is considered a Regionally Important Geological Site (RIGS) because a marker horizon, the Tollhouse Mudstone Bed, outcrops here which has provided stratigraphic correlation of the Clwyd Group limestones of Anglesey with those on the Great Orme (Waters et al., 2007).

The Leete Limestones pass up into the Loggerheads Limestone Formation (Asbian) which represents a shelf to deep shelf depositional environment. In Anglesey these strata are known as the Penmon Limestones, a sequence of muddy, bioturbated calcarenites, which are also known in the stone trade as the Penmon Marbles because of their ability to take a good polish and their attractive qualities subsequently. These are brown to grey mottled, bioturbated limestones, with abundant linear and branching burrows, and algal and peloidal grainstones rich in fossil of productid brachiopods and colonial and rugose corals. On the Great Orme, Loggerheads Formation limestones are white-weathering, reefal limestones. The overlying Cefn Mawr Limestone Formation (Asbian Brigantian) consists of predominantly argillaceous limestones with disarticulated thick-walled brachiopods and occasional thin cherty beds. The uppermost Red Wharf Limestone Formation (late Brigantian) is a return to massive, occasionally cross-bedded, sandy, fossiliferous and reefal grainstones and algal grainstones, once again representing a return to a deeper shelf depositional environment.

The equivalent strata on the mainland outcrop on the Vaynol Estate on the Menai Straits between Port Dinorwic and Bangor and have been locally divided into the Treborth and Dinorwic Formations, which correlate respectively to the Upper Loggerheads and Cefn Mawr Formations and the Red Wharf Bay Formation. The quarried units are on the Treborth Formation, and like their Anglesey equivalents, they are dominated by packstones and grainstones, though with a somewhat sparse fossil fauna, though crinoids are commonly encountered. Distinctively, these sandstones are stained pink and red due to the overlying continental red beds assigned to the Warwickshire Group (Greenly, 1928).

The Clwyd Limestones have been and still are widely quarried through their outcrop and are important regional building stones. Stone has been worked from the Penmon area since the Roman period. In addition to dimension stone, it should not be forgotten that lime production has also been a major driver for quarrying in the area. Huge lime kilns still exist *in situ* in Flagstaff Quarry. Another major use of Anglesey limestone was in industrial iron production where it was used as a flux. There is far less documentary evidence available concerning the ownership and leaseholders of the quarries than there is for the slate and granite quarries of North Wales, probably because these quarries were not connected by a railway network. Locating stone from building works and tracing it back through stone contractors to the quarries can therefore be problematic, even for major building projects such as Thomas

Telford's Menai Suspension Bridge. Thomas (2014) summarises what is known of the quarry business operating in NE Anglesey. In the 18th and 19th centuries, the land was owned by the Bulkeley Family who would have issued leases to quarrymen and quarry firms to work the quarries and limeworks. In the mid-19th century, Richard Kneeshaw and William Lupton, a partnership of quarry entrepreneurs who owned granite quarries at Penmaenmawr and on the Llyn Peninsula, were selling limestone in the region. A man called Edmund Spargo, a Liverpool mining consultant, occurs in the records for the later 19th and early 20th century. In 1909, Spargo was tasked by the Mersey Docks and Harbour Board to procure stone from the Penmon and Dinorben Quarries for the construction of harbour works (Thomas, 2014). This extensive use of materials brought about a revival of the quarries, and companies including the Dinmoor Quarry Ltd. and Tan Dinas Quarries Ltd. were established in the early 19th century. These companies were taken over in the mid-20th century by Cawood Wharton. Modern enterprises include Aber Quarry at Moelfre and Nant Newydd and Rhuddlan Bach Quarries in Benllech, all currently in local, private ownership (Cameron et al., 2020).

All facies of the limestone were used for building stone, and some were suitable as decorative stone (Figure 2.12). During the 19th century particularly, the brown-grey, bioturbated 'Penmon Marbles', found a market for the construction of chimneypieces and small-scale decorative architectural objects, as well as use in British *pietre dure* work and collectors' specimen marble tables. Black, bituminous limestones, again primarily used for chimneypieces, were quarried at Dinorben, on the Anglesey Coast, north of Moelfre (Watson, 1916; Daniell & Ayton, 1814). On the mainland, Loggerheads Limestone (locally called Treborth Limestone) was quarried on the Vaynol Estate and has been used in Caernarfon, Port Dinorwic and Bangor in the 18th and 19th centuries and also in repairs and renovations undertaken in Caernarfon Castle and the adjacent town walls. The pink colouration of the Vaynol limestones enables them to be differentiated from other local limestones when they are observed within the context of a building. The limestones were rarely exported outside North Wales, but Penmon Loggerheads Limestone was used to build the Town Hall in Birmingham in 1832, a building designed by the architects Hansom and Welch, who had previously worked at Beaumaris and were familiar with this stone. Quarries on the Great Orme at Llandudno were worked

Figure 2.12 Facies and appearance of Clwyd Group Limestones used as building stones. (a) Shaley Leete Limestone containing a fossil of a Productid brachiopod. The example, from Llanpeblig Church in Caernarfon, shows the rusty weathering crust. (b) Fossil colonial coral preserved in a block of Loggerheads Limestone in the Beaumaris Castle. (c) Cefn Mawr Limestone in Caernarfon Castle. A dark grey micritic limestone with sparse fossils and a pale grey weathering crust. (d) A rare example of a fossil goniatite in Cefn Mawr Limestone in the ashlar masonry of Plas Newydd. (e) 'Penmon Marble' facies of Loggerheads Limestone showing dense bioturbation. The stone here is used for dressings and decorative carving at the Penrhyn Hall in Bangor. (f) Porcellanous, white weathering 'Tandinas Limestone' in Llanpeblig Church, Caernarfon. The iron staining is due to an iron window casing immediately above this stonework. This unit is found at the top of the Leete Limestone.

from copper ores, and probably stone, during the Bronze Age (see Ixer, 2001) and then later for stone to supply the local towns of Llandudno and Conwy.

Outside the area of the Edwardian Castles and the region covered by this book, major quarries currently operate in the Clwyd Limestone Group at Halkyn and elsewhere in the Vale of Clwyd and Clwydian Hills for the production of lime and building stone up unto the present day. Here too, a few beds are worked from decorative stones such as the crinoidal Halkyn Marble, which is used locally for ecclesiastical ornaments and fonts (see Haycock, 2017). Active quarries operate in Moelfre, and piecemeal extraction of stone still occasionally occurs at Penmon for specialist and conservation work. Clwyd Limestone has seen a renaissance in recent years and is currently widely used in local and regional building works as well as for modern cement production. The Harbour (Yr Harbwr) complex of hotels and retail space in Caernarfon and the Deiniol Shopping Centre in Bangor are both recent builds in Clwyd Limestone. The Clwyd Limestone quarries are described in Thomas (2014), and Cameron et al. (2020) provide a list of currently operating quarries and their products.

2.8.3 Warwickshire Group

Upper Carboniferous sandstones are present in NW Wales but not as abundant as they are to the NE where thick deposits of Millstone Grit and Pennine Lower Coal Measures Groups are found. In Arfon, Upper Carboniferous sandstones were assigned, undifferentiated, to the Plas Brereton Formation of Howells et al. (1985); however, this is now considered an obsolete stratigraphic term, and the units described below are now officially ascribed as an undifferentiated formation within the Warwickshire Group strata of Bolsovian-Asturian age. Outcrops are restricted to a narrow strip between Caernarfon and Port Dinorwic on the mainland Menai Straits shoreline, bounded in outcrop to the south by the Aber Dinlle Fault of the Menai Straits Fault System. These strata have proven difficult to date and are spatially isolated from other outcrops of Warwickshire Group strata in the British Isles. Between Plas Menai and Y Felinheli, these red beds are composed of coarse breccias with clasts of predominantly felsite porphyries; rhyolitic-composition volcanic and volcaniclastic rocks which are well exposed on beach sections below Plas Menai and just east of Y Felinheli. Nearer Caernarfon at Port Waterloo, these units outcrop as red marls. These clastic sedimentary rocks were described in detail by Greenly (1938), who referred to them simply as the 'Menai red beds'. Warwickshire Group sediments are otherwise absent in North Wales west of the Denbighshire region and, indeed, the English Midlands where the type sections are recorded. This local facies with breccias rich in clasts of predominantly volcanic-derived rock fragments is unique to this region, and it is proposed here that the local formation name be retained. They are not important as building stones. The Plas Brereton Formation Breccias are used sparsely in rubble masonry at the 14th century church at Llanfair Is-Gaer at Plas Menai.

A quartz-arenite is frequently encountered as a building stone in Caernarfon which is possibly a member of the Warwickshire Group. This stone has not been observed in outcrop, and therefore it is assumed that it has been entirely quarried out. It is, nevertheless, a distinctive stone, stained purple-red to pink by iron; it exhibits well-developed and striking liesegang banding which makes this a rather attractive stone. This stone is only seen as a building stone in Caernarfon where it is used for a number of early 19th century domestic buildings in the area of the town centre. It is also used along the shore of the Menai Straits both west and east of the town for various structures. This stone is the main building stone used to construct the remains of the jetty, sea wall and wharf at Port Waterloo and the house of Plas Brereton and cottages on the Plas Brereton estate. This red sandstone is also used in field walls west and south of Caernarfon in the vicinity of Coed Helen, including for the construction of the lych gate at St Baglan's Church, which dates to the early 18th century.

A borehole was sunk in the grounds of Plas Brereton House, which is located 2 km north west of Caernarfon town centre on the shore of the Menai Straits, in 1903. Apparently drilled with the intention to prospect for coal, the core has been lost, but Greenly (1938) had access to the sedimentary log obtained in this operation. The borehole was drilled to a total depth 203 m and encountered a 4 m-thick bed of 'hard, red sandstones' at a depth of ~80 m, as well as numerous other hard and soft sandstone beds at various levels.[3]

---

3 The borehole penetrated the top of the Clwyd Group Sandstones and was abandoned, concluding that the Coal Measures were therefore missing from the stratigraphy.

It is not impossible that this or other sandstone beds outcrop on the surface further along strike to the south, perhaps in the vicinity or Port Waterloo or Doc Victoria.

Greenly (1928) also surveyed the area of Coed Helen, across the Seiont from and directly opposite to Caernarfon Castle for Carboniferous strata investigating the presence of an outlier of Lower Carboniferous rocks which are shown on the 1852 and 1851 British Geological Survey map (BGS, 1852). Greenly notes that there was neither extant strata, nor evidence of quarrying although there was a 'suspicion of the presence of Red Measures'. A low-lying, densely wooded area immediately south of the modern caravan site on Coed Helen may well have been a quarry at an earlier date. However, it is true that there is scant evidence of Carboniferous red beds, or indeed any other strata, outcropping in the area, and they do not feature on the most recent geological map of the region (British Geological Survey, 2015).

This red sandstone is an important building stone is Caernarfon and is hereby tentatively assigned to the Warwickshire Group and as a building stone, will be referred to herein as the Plas Brereton Sandstone (Figure 2.13).

Figure 2.13 Typical appearance of the pink liesegang-banded sandstone named here as the Plas Brereton Sandstone. The precise provenance of this stone is unknown, but it may belong to the Warwickshire Group of Upper Carboniferous continental sandstones. It is used widely for building in Caernarfon in the 18th and 19th centuries.

2.8.4 Gloddaeth Purple Sandstone

Outcropping in a limited region north of the railway town of Llandudno Junction on the Creuddyn Peninsula, the Gloddaeth Purple Sandstone is of uncertain age, but it has been assigned to the Warwickshire Group on the basis of (unpublished) palynological analyses (see Davies et al., 2011; Lott, 2010). It outcrops in the core of the Gloddaeth Syncline where it unconformably overlies the limestones of the Clwyd Group. This is a medium- to very coarse-grained gritstone with well-developed cross- and channel-bedding. It varies in colour across beds and cross-beds from white to red and purple shades and is often mottled in these colours (Figure 2.14). This stone does not outcrop well, but an ancient quarry exists in the grounds of Bodysgallen Hall which supplied stone in the Medieval period to nearby Conwy where it was used in the Castle and also at St Mary's Parish Church. It is also found in a surviving 14th century merchant's house (Aberconwy House) indicating its use continued into the later Medieval period. This is an important building stone of this period despite its being only used at Conwy and in the Conway Valley region. Gloddaeth Sandstone is used at Dolwyddelan Castle and St Mary's Church at Caerhun. St Mary's Caerhun is built on the site of the Roman Fort of Canovium, and one might speculate (albeit with little supporting evidence) that

Figure 2.14 Gloddaeth Purple Sandstone used as window dressings in Conwy Castle.

this stone had been quarried in the Roman period and reused in the church. The Gloddaeth Purple Sandstone was almost entirely worked out from its one and only quarry site probably as early as the start of the 15th century. This has meant that it has been absent on geological maps and has only relatively recently been recognised as an important part of the local stratigraphy. The lack of knowledge of the existence of this stone has meant that previous workers have interpreted its occurrence in various buildings in the Conwy Valley as imported Permian sandstone from Cheshire, which though finer-grained, also exhibits mottled facies. Such misattributions and confusion are frequently encountered in the literature referring to the architecture of the Conwy Valley, and the stone requires direct observation to enable secure identification and differentiation from other red and white mottled sandstones (which arguably includes the 'peaches and cream' stone of Davies (2018a), itself also quarried out, from the Basement Beds facies of Moel-y-don on Anglesey). However, it is likely that the use of this stone will be seen to be more widespread.

2.9   INTRUSIVE IGNEOUS ROCKS

The main intrusive igneous complex on Anglesey, the Coedana Granite, is Neoproterozoic, as is a small stock on the mainland at Twthill in Caernarfon. Intrusions of similar age also occur in the SW Llyn Peninsula which is outside the region of study. However, there are a large number of small igneous intrusions, ranging in size and field relations from dykes, sills, stocks and small plutons in the area of mainland NW Wales. These magmatic rocks are coincident with the Ordovician phases of volcanic activity and intrude Ordovician and older strata. Compositionally they range from dolerites, which are very widely distributed in the region, through granodiorites, granites and rocks of more unusual geochemistry including a suite of quartz-latites. Many of these intrusions, and especially those near the coast, have been quarried at some time over the last two centuries, even if only for trial pits. Their main use was not as dimension stone (although some lithologies have found local use of these rocks as building stones) but as paving setts and kerb stones as well as roadstone and aggregate. A boom for these materials occurred as a consequence of the growth of cities of the industrial revolution which expanded on an unprecedented scale throughout the 19th century, particularly Liverpool, Manchester and Birmingham. These stones were also used locally for paving and for engineering works such as The Cob at Porthmadog. Requirement for paving stone should not be underestimated. Enormous quantities of granite for paving and roadstone were consumed at this time. A major British centre for production of such materials was located in Leicestershire where the Mountsorrel, Markfield, Groby and many other intrusions were being quarried. Mountsorrel remains one of the largest quarries in Europe, and it was worked almost entirely for paving setts during the 19th century. Leicestershire's dominance in this market had an influence in North Wales and Leicestershire granite companies had stakes in and ownership of a number of quarries in the region, and also quarry workers from Leicestershire were recruited as quarrymen. The demand for paving setts meant that many granites which had previously been seen as having little value as dimension stone because they had too closely spaced joint planes suddenly had huge economic value as setts and cobblestones. The granite 'bubble' of North Wales made a few people very rich over a short time, until road building moved on and setts became replaced by tarmac. Even then roadstones, particularly dolerites, remained in demand and a few NW Welsh quarries continue to supply this commodity. A more recent requirement of large blocks of granite rubble is as sea defences in these times of rising sea-level and increased storminess. Many granite quarries worldwide have found there otherwise waste rock to be a lucrative commodity in this environment. Both Trefor Granite and Nanhoron Granite, both on the Llyn Peninsula, are worked to produce sea defence stone.

2.9.1 The Neoproterozoic granites: Twthill and Coedana Granites

Outcropping along the NNE-SSW trending Menai Straits Fault System, the Twthill Granite forms in outcrop an elongated lens-shaped stock on the mainland between Caernarfon and Port Dinorwic (Y Felinheli), most of its outcrop is largely obscured by superficial deposits. However, it is well exposed in the town of Caernarfon where it forms a small but prominent rocky hill with evidence of quarrying on the NE side (Greenly, 1944a, 1945; Bonney & Houghton, 1879). Here the granite is coarse-grained, and a pale-orange shade overall, being composed of orange-pink feldspar and white milky quartz. Grain size decreases considerably westwards towards Y Felinheli. Though generally ascribed to the Ordovician, it has recently been assigned a late Proterozoic date of 615.2 ± 1.3 Ma by Schofield et al. (2008). Twthill Granite is used mainly for walling, especially in the eastern end of the town of Caernarfon.

The Coedana Granite intrudes high-grade, quartzo-feldspathic gneisses in central Anglesey which have been dated to 666 Ma (Horák et al., 1996; Asanuma et al., 2017). The Coedana granite is coarse-grained and variably porphyritic, composed of pink K-feldspar, quartz and muscovite and is highly fractured and sheared, which has developed a foliation. Tucker and Pharoah (1991) have dated the Coedana Granite to 613 ± 4 Ma (U-Pb zircon). This is a very weathered granite and although field stones were exploited during the Neolithic and used locally for dry stone walls, it is not suitable for use in construction beyond use as an aggregate. The Coedana Granite is actively quarried for aggregate at Gwalchmai and the quarries have been in operation since the 1950s. However, the fractured nature of the rock has not made it viable as a dimension stone (Jana Horák, *Personal Communication*).

2.9.2   Penmaenmawr Granodiorite

The Penmaenmawr Granodiorite is an upper Ordovician intrusion located on the hills of Penmaenmawr and Graig Lwyd on the Coast between Bangor and Conwy. It is the largest quarry in igneous rocks in NW Wales and has been worked since the Neolithic. It is an important source of local and regional stone and therefore considered here separately. The history of historical period quarrying has been documented by former quarry employee, Ivor Davies (1974), and is well summarised in Thomas (2014). The Penmaenmawr Granodiorite is an intermediate intrusion of Upper Ordovician Age emplaced into Nant Ffrancon Group sedimentary rocks. Geochemically the pluton has calc-alkaline affinity, with significant quartz content throughout its outcrop (Tremlett, 1997). For the most part, it is a fine-grained, grey to grey green-coloured rock with a sub-conchoidal fracture, composed of plagioclase, orthopyroxene, clinopyroxene, quartz, orthoclase and biotite (the latter chloritised). Secondary epidote and prehnite are also present in patches. Grain size can be variable from aphanitic to medium-grained, even pegmatitic green and white 'brindled' varieties. When fine-grained, the granodiorite is grey in colour.

Screes from Penmaenmawr and Graig Lwyd were worked in the Neolithic for the production of hand axes, and the area is regarded as an internationally important lithic production site for this period (Kenney, 2017; Hazzledine Warren, 1919 and see Chapter 3). Two main areas of historical quarrying were originally located on Graig Lwyd, the site of the Neolithic axe factory, and on Penmaenmawr Mountain, these latter quarry sites were known as Gorllewin and Dwyrian. The Penmaenmawr Mountain quarries were opened in the 1830s by the Brundrit and Whiteway Company. Philip Whiteway was a Cheshire geologist, and the company he established with Bundrit maintained ownership of the quarries up until 1911. Whiteway had previously had stakes in quarries in the Runcorn area of Cheshire. Works at Graig Lwyd (the old quarry, Hen Chwarel) were opened four years later by Liverpool-based railway contractor Thomas Brassey and his business partner John Tomkinson, but they soon gave up their lease to three other Liverpudlians, Richard Kneeshaw, J. T. Raynes and William Lupton. The Darbishire family bought up the leases and took over Hen Chwarel in the 1870s and in 1911, amalgamated all the quarries at Penmaenmawr with the works at Trefor (see below) as the Penmaenmawr and Welsh Granite Company. The quarry continues working today, albeit on a very small scale supplying railway ballast[4] and has been operated by Hanson Aggregates since 1995 (Thomas, 2014; Cameron et al., 2020; Davies, 1974).

In addition to roadstone, railway ballast and paving setts, the Penmaenmawr Granodiorites are widely used stones throughout North Wales especially in the construction of civic buildings and particularly for churches and chapels, during the second half of the 19th century. Most of the buildings of this period in the quarry towns of Penmaenmawr and neighbouring Llanfairfechan are built from this stone, and it is used in Caernarfon, Bangor and Conwy. Typically, Penmaenmawr Granodiorite is used in random and coursed rubble masonry styles, using angular, polygonal, knapped blocks of around 20 cm maximum dimension. It was also used for paving setts and cobbles, and these were the main money-making commodity and were exported widely throughout the British Isles.

---

4 Not to be confused with ships' ballast, railway ballast is loose aggregate, used as the foundation to lay railway tracks.

Penmaenmawr Granodiorite was certainly exported to Dublin, where its distinctive brindled facies is used for paving in the courtyard of Trinity College, and it was also exported to the growing cities of Northern England and the Midlands.

2.9.3   The intrusive igneous rocks of Snowdonia

A large number of relatively small igneous intrusions are encountered throughout Snowdonia. The two largest are the Mynydd Mawr Pluton, a riebeckite-bearing granite (which is not quarried), and the Tan-y-Grisiau Granite, located just to the south of Blaenau-Ffestiniog, which was quarried in the early 20th century. Also in the vicinity of Blaenau-Ffestiniog area are a suite of much under-studied, intrusive peralkaline rocks known as the quartz-latites (Bromley, 1965). These have been quarried at Manod near Blaenau-Ffestiniog and at Arenig Fawr, in the remote uplands between Blaenau-Ffestiniog and Bala. To the south around Porthmadog are a suite of dolerite sills which intrude the shales and siltstones of the early Ordovician Tremadocian Dol-cyn-afon Formation. Dolerite sills of a similar age have also been worked on a limited scale in the Conwy Valley between Trefriw and Gwydyr Forest at the small quarries of Ty'r Mawn and Pant-y-Carw (Thomas, 2014). Dolerites were also quarried on a small scale at numerous sites on the Llyn Peninsula. Young et al. (2002) note small quarries here at Pen-y-Gaer, Nant-y-Carw and Dinas. A dolerite dyke also cuts the Garnfor Complex and is quarried at Trefor.

Thick Ordovician dolerite sills were quarried at Moel-y-Gest above the town of Porthmadog and on the east side of the Glaslyn Estuary at Minffordd at Penrhyndeudraeth. These are fine to medium-grained, greenish-coloured dolerites, with grain size visible to the naked eye (they are often referred to as microgabbros). The quarry at Minffordd near Penrhyndeudraeth is generally known today as Garth Quarry. It was working stone as early as 1811 to supply material for the construction of The Cob barrier across the Glaslyn Estuary, linking Porthmadog and Penrhyndeudraeth. Continual modifications and enlargement of The Cob, especially with the building of the Ffestiniog Railway in the 1830s and the later introduction of steam locomotives on the line in the 1850s, brought the stone greater value and the quarry expanded. In the second half of the 19th century, it was owned by the Cambrian

Granite Company, who also owned quarries of the Llyn Peninsula (see below and Thomas, 2014). Garth Quarry continues to operate today for roadstone production (Cameron et al., 2020; Jones & Hankinson, 2016). Dolerites from both Garth and Moel-y-Gest, above the town of Porthmadog, have been used as building stone in the local towns in the mid to later 19th century. A particular use of this stone was as angular rubble, pressed into concrete and used to create a weather-proof façade (Figure 2.15).

Quartz-latite was quarried at the Manod Granite Quarry at Manod, near Blaenau-Ffestiniog. The Manod Intrusion is emplaced into the slates of the Nant Ffrancon Group. Igneous rock from Manod was transported from the quarry using the Ffestiniog and Blaenau branch line of the Ffestiniog Railway. This stone was worked for railway ballast and paving setts from the later 19th century and was in operation for roadstone from the 1950s up until 1972, run by Cawood Wharton & Co. Ltd. A similar stone was quarried in the more remote Arenig Quarry, which was served by its own railway. This is an enormous quarry, opened in 1908, and its main purpose was for the production of railway track ballast for the construction of the (now defunct) railway line between

Figure 2.15 Angular cobbles of Minffordd Dolerite pressed into concrete and used in the late 19th century as a surface effect on buildings in Porthmadog. The largest cobbles are ~8 to 10 cm diameter.

Ffestiniog and Porthmadog to Bala. The quarry itself was located on the railway and had its own station. The railway and the quarry closed in 1960. The Manod and Arenig quartz-latites are intrusive porphyritic rocks which were once thought to be extrusive andesites. Texturally they have white plagioclase phenocrysts set in a grey, aphanitic matrix (Bromley, 1965). Little is documented about the geology of quarrying of these stones (although there is considerable documentation concerning the accompanying railways; see for example Southern, 1995) and their petrology, history, and the subsequent use of the quarried stone requires further research.

The Tan-y-Grisiau Granite is quarried at the Groby Quarry at Tan-y-Grisiau, which had its own branch line on the Ffestiniog Railway. This granite is fine-grained and grey-green in colour, composed of plagioclase, quartz, perthite and chlorite replacing biotite (Smith, 2007). The quarry was opened in 1901 and operated for some 30 years. Tan-y-Grisiau Granite was used for paving setts as well as roadstone. It was owned by the Leicestershire-based quarry firm, the Groby Granite Quarry, and there is another, much larger Groby Quarry in the eponymous village in Leicestershire which works the Neoproterozoic South Charnwood Diorites. These units were also worked as setts and roadstone. Groby Quarry, also known as Ffestiniog Quarry, is now preserved as it exposes the roof zone of the granite, which is extensively mineralised and the overlying hornfels facies of the Dol-cyn-afon slates host rock. Geophysical data suggest a far more extensive igneous mass at depth (Smith, 2007).

2.9.4 The Nefyn Cluster and Nanhoron Intrusions of the Llyn Peninsula

A number of stocks and small plutons were intruded, mostly at a high level into the crust, on the Llyn Peninsula during the Caradoc stage of the Ordovician (Tremlett, 1962). These can be subdivided into the Nefyn Cluster Intrusions in the north and eastern Llyn and the Nanhoron Suite in the central Llyn Peninsula. These igneous rocks are predominantly peralkaline and range in composition from dolerites, through diorites, granodiorites, microgranites and rhyolitic felsites. Each intrusion has distinctive textures and chemistry, and many have been quarried at some point in history, and some are important local and regional and even international building and paving materials. Long prior to any quarrying activity, eroded and frost shattered screes would have provided readily available building materials for Iron Age hillforts, the castle at Criccieth and as rubble and ashlar masonry in churches and domestic buildings. In the 19th century, the rapidly expanding conurbations of Manchester, Liverpool and Birmingham, hubs of the industrial revolution, had a seemingly insatiable requirement for kerb stones, paving setts and cobblestones. Many small quarry companies sprang up, hoping to capitalise on this need for non-slip paving stone, and inevitably some operations were more profitable than others. Several of these operations were set up by the quarry entrepreneur Samuel Holland. We have already encountered Samuel Holland junior working at the Oakeley Quarries in Blaenau-Ffestiniog in the early 19th century. He opened the quarries at Trefor and at Gwylwyr on the Llyn Peninsula, the latter to provide paving setts for Manchester. Holland sold these quarries as soon as they became profitable, to stone contractors intent on supplying building materials to the new industrial towns of the North of England and desirous to cash in on the North Welsh granite 'bubble'.

The largest quarry on the Llyn Peninsula and the most important in terms of building stone is at Trefor. Here, the Garnfor Multiple Intrusion comprises a suite of three granitoids, technically granodioritic in overall composition, a grey, vitreous porphyry from the 'Inner Intrusion', a pink variety from the 'Outer Intrusion' and a stock of hypersthene-bearing, blue-grey porphyry (the 'Blue Rock'). In the majority, the Trefor rocks are porphyritic with plagioclase and pyroxenes with variable amounts of quartz (Cattermole & Romano, 1981), and all rock types have been quarried. Trefor Quarry, known by everyone locally as Y Gwaith Mawr (the big works), was opened in the 1830s by Samuel Holland primarily to supply quarry setts for road paving. The village that grew up around the quarry and its wharf was named after the quarry foreman Trevor Jones, who despite his local-sounding name was a Leicestershire man, employed by Holland for his experience of quarrying granite for paving setts (Thomas, 2014). Holland moved on to pursue his career in politics, and the Trefor quarry business became the Welsh Granite Company which also owned and operated other, smaller quarries on the Llyn Peninsula. Horse-drawn tramways were built to carry stone from the quarries down to the wharfs on the coast and a steam railway was installed in the 1870s. By the end of the 19th century, Welsh Granite were the largest producer of paving setts in Wales. There had always been connections between the quarries at Penmaenmawr and those at Trefor. Holland has employed experience granite workers at Penmaenmawr to assess the quality of the stone at Trefor, and then in 1911, the Darbishire Family bought the Trefor Quarries, and they became part of the new Penmaenmawr and Welsh Granite Company. The Trefor Quarries went into decline in the early 20th century as the demand for paving setts declined (Bendall, 2004). Trefor Quarries were bought back into private ownership in 1985 by, the appropriately named, R. Trefor Davies, a descendant of a family of quarrymen who had worked in granite quarries on the Llyn Peninsula (Thomas, 2014). Despite its fame for producing setts, the Garnfor Intrusions have well-spaced joints and large blocks of up to $8\,m^3$ can be extracted from Trefor Quarry meaning it has also found use as dimension stone and armour stone. A single, 8 tonne block of Trefor Blue Granite has been used for a memorial to Owain Glyndwr erected in Corwen in 2007, and a 1.5 tonne block of undressed stone marks the site where Llywelyn ap Gruffydd was killed in 1282 at Cilmeri in Powys. A polished slab of Trefor Granite was taken to the Falkland Islands as a memorial to the HMS Glamorgan. The stone has also been used for memorials in London and much more locally as the plinth to the statue of David Lloyd George in Caernarfon (Figure 2.16). Trefor Granite is now quarried to produce stone for sea defence structures, dimension stone and also curling

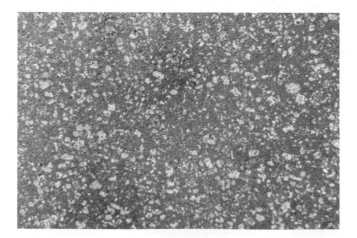

Figure 2.16 Trefor Porphyry on the plinth of the statue of David Lloyd George in Caernarfon. The field of view is 20 cm.

stones. Trefor curling stones were used as competition stones in the 2002 Salt Lake City Winter Olympics (Bendall, 2004; Elis-Gruffydd, 2008; Thomas, 2014). Trefor Granodiorite has been used as dimension stone in the church at nearby Llanaelhearn and in several churches in Caernarfon and elsewhere. It is used for sea defence blocks and building the sea wall at Dinas Dinlle.

Felsite plutons, composed of fine-grained, variably porphyritic intrusive igneous rocks are a common component of the Nefyn Cluster Intrusions. Pink weathering, very distinctive porphyries, with white feldspar phenocrysts set in a grey aphanitic matrix are quarried from the Mynydd Nefyn Pluton at Gwylwyr Quarry and also at Tan-y-Graig and Tyddyn Hywell in the Gyrn Ddu Microgranite at Clynnog Fawr. The Gyrn Ddu Microgranite had also been exploited locally for rubble and coursed rubble masonry since the early Medieval period, and indeed scree was used in the Early Christian Period (5th–9th centuries CE) for making memorial stones (inscribed stones) and later in the 15th century for building St Beuno's Church at Clynnog-Fawr (Figure 2.17). The Tan-y-Graig and Tyddyn Hywell quarries were worked in the first few decades of the 20th century by the Kidderminster-based

Figure 2.17 Gyrn Ddu Microgranite used at St Beuno's Church in Clynnog Fawr. Note the pink weathering crust which develops on the otherwise grey porphyry. White feldspar phenocrysts are clearly visible.

British Grey Granite Company and the Leicestershire firm, the Enderby and Stoney Stanton Granite Company. A very similar stone was quarried from the Gwylwyr Quarry in the Mynydd Nefyn Pluton where quarries were opened in the 1830s by Samuel Holland and which later came under the general ownership of the Trefor Quarries and the Welsh Granite Company. Quarrying at Gwylwyr continued up until the end of World War II (Thomas, 2014). At Criccieth, a well-jointed felsite stock has provided both a dominating, defensive site and the main building stone for the rubble masonry construction of the castle. This is an aphanitic, pale pink-coloured rock which was quarried from the Castle Rock and from the neighbouring outcrop of Dinas. This stone has not been worked more recently that the 13th century.

A very distinctive stone from the Nefyn Cluster Intrusions is the Bodeilas Granodiorite which was quarried at Penrhyn Bodeilas from a small stock adjacent to the Mynydd Nefyn Pluton. This is a grey green, coarse-grained plutonic igneous rock with centimetre diameter clots of ferromagnesian minerals. St Baglan's Church at Llanfaglan near Caernarfon was constructed of this stone in the 17th century which suggests much earlier workings on this site (Figure 2.17). Penrhyn Bodeilas was further developed and worked for setts and cobbles in the mid-19th century again in association with Samuel Holland and later the Welsh Granite Company. Penrhyn Bodeilas is a small headland, and stone could be quarried from coastal exposures and shipped directly from the site. In the 19th century, wharfs and jetties were constructed for this purpose.

The Iron Age hillfort at Carreg-y-Lam in Nant Gwrtheyrn was partially destroyed by quarrying for felsites of the Carreg-y-Llam & Porth-y-Nant Microgranites. Three quarries work the stones in this area, Caer Nant (The Rivals Quarry), Nant and Carreg-y-Lam. Both intrusions are composed of a grey or buff, porphyritic crystalline rock with feldspar phenocrysts up to 2 mm long and specks of chlorite and minor clinopyroxene. The quarries were established in the 1850s for the production of paving setts and kerbs. Caer Nant Quarry[5] went through a number of private owners before being acquired by the Cambrian Granite Company in the late 1870s.

---

5 Caer Nant was also known as The Rivals Quarry, not because of any strife between quarry owners in this small area but as a corruption of the Welsh name for the mountain above the site, Yr Eifl.

The Caer Nant quarrying business continued to change hands several times during the later 19th century and was owned by The Rivals Granite Ltd. in 1904 and later taken over by the Croft Granite, Brick and Concrete Company, another Leicestershire firm, in 1922 (Thomas, 2014) (Figure 2.18).

Nant Quarry was opened by Anglesey entrepreneur Hugh Owen in 1951. It too changed hands several times over the next decade until coming into the possession of Kneeshaw and Lupton in 1861, who had previously owned quarries at Graig Lwyd on Penmaenmawr Mountain. They created the Port Nant Quarries Ltd. As well as production of setts, the stone was also used to build the new village to house the quarry workers at Port-y-Nant, many of whom were migrants from Leicestershire. Port-y-Nant was constructed between 1875 and 1878. Production peaked at Nant in the 1880s and continued until 1922 when a storm destroyed the jetties and wharfs. The cottages and shop of the Port-y-Nant 'model village' are now restored and preserved, and this is the home of the National Welsh Language Centre (Malaws, 2006).

The largest quarry in Nant Gwrtheyrn, Carreg-y-Lam Quarry was opened in 1908 by H. J. Wright, who had previously run the

Figure 2.18 The distinctive, coarse-grained, igneous texture of Penrhyn Bodeilas Granite, speckled with clots of ferromagnesian minerals.

Caer Nant Quarry and before that had involvement in various slate and gold mining stakes in North Wales. This new quarry had road as well as sea access and became a modern, going concern as Carreg-y-Lam Ltd. in the 1920s. In 1934 it became part of National Amalgamated Roadstone group of companies and continued to operate until 1963 (Thomas, 2014).

The Liverpool and Pwllheli Granite Company was established in 1858 (Horák, 2020) and worked the Carreg-yr-Imbill microdolerite stock, a pipe-shaped intrusion at Gimblet Rock, once a prominent island just offshore of the town of Pwllheli. Again primarily quarried for kerbstones and setts for export, this stone was also used as a dimension stone for the main civic buildings in Pwllheli, including the Town Hall and a number of chapels and of course, it was also used for kerbstones. It is a particularly striking and distinctive stone, overall black in colour and composed of plagioclase, clinopyroxene and minor olivine. It is predominantly medium-grained but containing distinctive pipes and veins of plagioclase-pyroxene micropegmatite. The outcrop was almost entirely quarried out, and the quarry site on the now non-existent Gimblet Rock has been landfilled and redeveloped as a holiday park.

In the south-central Llyn Peninsula, many of the Nanhoron Suite of granitoids have been quarried at some time and to some degree in the past, and quarrying for aggregate and sea defence blocks is still current at Nanhoron where the lithology is a granophyric microgranite. Also part of the Nanhoron Suite, the Llanbedrog Granophyre is a yellow to grey felsite with fined grained matrix and 5 mm phenocrysts of feldspar, quartz and green biotite scattered throughout. Like the Nanhoron granophyre, it exhibits a granophyric texture on a microscopic scale. A number of quarries are located along the shoreline on the south and east margins of the Llanbedrog stock, which forms a prominent hill immediately south of the village of Llanbedrog (Young & Gibbons, 2007). The quarries lie within the Madryn Estate, which was owned by Sir Love Jones Parry (1781–1853) and then his son Sir Thomas Duncombe Love Jones Parry (1832–1891).[6] The Estate issued a number of leases for quarrying the site in the 1870s and quarrying at Llanbedrog continued, under the auspices of the Cambrian Granite Company until the 1950s. Samples given to John Watson in the early 20th

6 Jones Parry left North Wales in 1862 with Lewis Jones of Caernarfon, to found the Welsh settlement in Patagonia, Argentina, which was named Puerto Madryn after the Llanbedrog Estate.

century for his building stone collection housed the Sedgwick Museum, University of Cambridge, were supplied by the Clee Hill Granite Company, who presumably owned the quarry at the time (Watson, 1911). The main use of the stone was once again setts and kerbs, but the Llanbedrog Granophyre was used as a dimension stone in the construction of sea defences (Hyslop & Lott, 2007) and most notably at the manor house and now art gallery of Plas Glyn-y-Weddw in Llanbedrog. This mansion was built in 1857 for Sir Love's widow, Elizabeth Parry Jones (Haslam et al., 2009). Associated acid volcaniclastic rocks from the Caradoc-age Llanbedrog Volcanic Group (Rushton & Howells, 1998) have been used for local wall building, and these were used in the construction of ancient monuments. Columnar jointed pyroclastic deposits from this unit were exploited for memorial posts in the Early Christian period; examples are The Penprys Stones which are now on display in the porch at Plas Glyn-y-Weddw.

2.10   PLEISTOCENE AND QUATERNARY DEPOSITS

Mesozoic and Cenozoic deposits are absent from the area, with the exception of Tertiary fluvial sediments, known to exist underneath superficial deposits to the west of the Mochras Fault at Harlech. During this time, North Wales and the west of the British Isles as a whole were undergoing regional tectonic uplift as a response to the opening of the North Atlantic Ocean. In the Quaternary, the resulting highlands were intensely glaciated during the Pleistocene Ice Ages leaving behind a textbook example of a glaciated landscape in the mountains of Snowdonia (Hughes, 2009). The Arfon region is covered in extensive deposits of diamicton tills, moraine and fluvial glacial deposits, a deformed sequence of which is well exposed in cliff sections at Dinas Dinlle Beach, 5 km south of Caernarfon.

   These superficial deposits should not be overlooked as an important local source of building materials, in terms of both stone and clay for brickmaking. Coarse-grained deposits contain abundant rounded cobbles and boulders of lithologies representing the Ordovician hinterland of Snowdonia and are dominated by distinctive volcanic and volcaniclastic rocks often showing well-developed igneous textures typical of the welded tuffs and other volcanic and volcaniclastic rocks of the Snowdonia volcanic centres. Porphyritic igneous rocks, granites, gneisses, quartzites, sandstones and slate are also common clast components. These boulders are used to form rubble masonry wall cores at Segontium Roman

Fort and at Caernarfon Castle, and they were widely used, roughly shaped into ashlars, in vernacular buildings and in field and garden walls (Figure 2.20). Boulder clays are been actively quarried at Bryncir near Garndolbenmaen for gravels.

Clay-rich diamictons have been exploited in the town of Caernarfon for tile and brick making since the Roman period, but more actively over the last two centuries. The two main quarries are at Peblig and Seiont in the valley of the River Seiont, with other brick pits located nearby at Griffiths Crossing and Llanfaglan. A Roman tile kiln, part of a potters workshop which would have exploited local clays, was identified during road-widening works in the 1970s (White, 1985). Industrial-scale clay quarrying at Caernarfon, with onsite firing facilities began in c. 1850. The bricks were of high quality but were only used locally and were not used for decorative work and as facing bricks. Production continued until the 1960s, but the Seiont Quarries have been worked on and off to this day. A history of the brickworks is provided in Harris Jones (2000) and Roberts (2007). Caernarfon bricks are both red and yellow-coloured, the colour is dependent on the local iron content of the clays and firing conditions. They were used locally but seen as inferior quality and less popular with architects and the general public than the high-quality, bright red bricks brought in from Ruabon in NE Wales (Figure 2.19).

Figure 2.19 Cottages in Criccieth in the eastern Llyn Peninsula, built from roughly dressed, glacial boulders.

2.11   STONE FROM OUTSIDE THE REGION

2.11.1   Merionethshire Slate and other Welsh slates

Good-quality slates of Late Ordovician age also outcrop from near Tywyn on the coast south of the Harlech Dome in a belt that runs through Corris and Aberllefenni to Dinas Mawddwy, on the northern side of the Aberdyfi Valley. These slate-bearing strata of the Abercwmeiddaw Group and Abercorris Group are of latest Ordovician, Hirnantian age. The unit snakes across north central Wales with a broadly SW-NE Strike, dipping steeply to the SE. A series of folds deform the unit with ~ NW-SE trending axes, giving a zig-zag outcrop pattern. The cleavage is parallel to the strike and vertical. The marker horizon which delimits the base of the slate-bearing strata is a unit called by quarrymen Nod Glas (blue mark) which is a soot-black, pyritiferous, graptolite-bearing shale. Above this is the Broad Vein Slate, subsequently overlain by the Red Vein (rusty weathering) at the base of the Abercorris Group, then the thin Narrow Vein. The upper limit of the Narrow Vein terminates against an abrupt change to the grits of the Garnedd Wen Beds of the Abercorris Group.

The Narrow Vein (Abercorris Group) produced particularly high-quality slate, but the cleavage is variably developed, so that as well as roofing slates, Aberllefenni produced cut slabs of greenish-coloured slate used for counter tops, paving and windowsills. Narrow Vein slate took a relatively good polish. Davies (1880) describes the position of the Veins in the quarries, and Pugh (1923) more than adequately describes the local geology and provides an excellent map of the Corris area.

Slate was also produced in a number of places in South Wales including Gilfach in Pembrokeshire. Here metamorphosed, Arenigaged, volcaniclastic sandstones of the Foel Tyrch Formation were quarried at Llangolman in Pembrokeshire from the 16th century up until the 1980s. These thick, green-coloured slates were preferred over the local Cambrian slate by the architect of Bangor University, Henry Hare and used for its roof in 1911.

2.11.2   Carboniferous and Permo-Triassic Sandstones

The region of NW Wales is relatively short on good-quality sandstone freestones of the type used for quoins, window dressings and tracery and for general use in the ashlar masonry styles which were popular in 19th and early 20th century British civic architecture. Despite the good quality of the coarse-grained, Lower Carboniferous Basement Beds and their widespread use, these stones could not be cleanly dressed and were therefore not suitable for the construction of fine ashlar masonry. From the Roman period onwards, stone had been imported to Arfon by sea from North East Wales and the English Cheshire Basin where a well-developed sequence of Upper Carboniferous and Permo-Triassic sandstones have been quarried since Antiquity. Two Upper Carboniferous, fluvial-deltaic sandstones that are important in North Welsh regional architecture are Cefn Stone and Gwespyr Sandstone (Haycock, 2018; Lott, 2009; Roberts, 2009). The owners of the Cefn Quarries in Denbighshire had connections with the Oakeley Family who owned the slate quarries in Blaenau Ffestiniog, and as a consequence, a great deal of this stone has been used in the region. It was also popular as a building stone in cities such as Liverpool, Birmingham and Manchester.

2.11.2.1   Millstone Grit

Upper Carboniferous sandstones are the most important construction materials in north central and north-west England and are widely quarried across the counties of Lancashire, West Yorkshire and Derbyshire. The Namurian sandstones of the Millstone Grit Group outcrop widely in this area and, as the name suggests, were gritty, coarse-grained, fluvio-deltaic sandstones which were used, amongst other things, for the manufacture of millstones, which were exported throughout the British Isles. Many of these sandstones were also exploited as building stones and have many local quarry names which have been incorporated into the stratigraphic nomenclature of the region. These include important building stone members including the Pendle Grit, Chatsworth Grit, Ashover Grit, Haslingden Flags and the Rough Rock, the latter is the uppermost sandstone in the sequence. The Millstone Grit Group attains a thickness of almost 1,300m in Wharfedale. For the most part, the sandstones are lithic arkoses, which vary in grain size from very fine, massive freestones, to very coarse-grained, cross-bedded gritstones. They are interbedded with minor mudstones and coals. They were imported into NW Wales region in the 19th century and used for civic and religious buildings, particularly as dressings used on otherwise brick-built structures.

In North Wales, an important formation of the Millstone Grit, with respect to building stone is Gwespyr Sandstone. This is a predominantly buff coloured, medium- to fine-grained arkosic sandstone. It is bedded and cross-bedded on a variety of scales which weather out to form tafoni. Gwespyr Stone was quarried near Talacre in Flintshire and at Berwig near Wrexham. Gwespyr sandstone is an important freestone used mainly in North Wales since at least the 8th century AD (see Haycock, 2018), and it was widely used in the 19th and early 20th centuries. According to Roberts (2009), Gwespyr Stone was used for Basingwerk Abbey, Flint Castle, St Asaph Cathedral, parts of Rhuddlan Castle, and it was also considered as a building stone for the Houses of Parliament in London. Gwespyr Sandstone is used in 19th century renovation work at Caernarfon Castle.

2.11.2.2   York Stone and other Coal Measures sandstones

The lithologies known generically as York Stone[7] are flaggy and massive sandstone members from the Pennine Lower Coal Measures Group (PLCM) which are very widely quarried across south Lancashire and Yorkshire. The stratigraphic group is composed of alternating layers of sandstone, siltstones and coals and is of Westphalian age and overlies the Millstone Grit. Many of the sandstones were quarried and are designated as named members, which can be regionally and locally distinguished (see Lott, 2012; Geldard et al., 2011). The presence of the fossiliferous so-called 'Marine Bands' demonstrate short-lived marine incursions in the lower sections of the PLCM, but the middle and upper sections are terrestrial with a fluvial to deltaic origin and, as the name suggests, important coal-bearing strata in the United Kingdom. Another substantial group of clastic sediments, the PLCM unit is up to 750 m thick in Lancashire. Flaggy York Stone has been the *de facto*

---

7 The name York Stone has become standard, but it is misleading. The stone has no association with the City of York or its environs, which sit on Mesozoic strata. To further confuse things, flagstones from the Millstone Grit Group (i.e. Haslingden Flags) are also referred to as York Stone.

paving stone used in England and Wales since the 18th century. It is used for pavements in Conwy, Caernarfon and Beaumaris as well as many other towns and villages in North Wales. It is also used, polished, in the interior of Penrhyn Castle. York Stone is derived from various sandstone members within the PLCM; i.e. the Elland Flags, Pennistone Flags, Greenmoor Rock and many others. The Elland Flags are the most widely exploited and are worked at a number of quarries across west Yorkshire and Lancashire. At the type area of Elland, near Halifax, the Elland Flags have been worked since the 12th century. By 1900, there were at least forty flagstone quarries in operation. The industry went into decline after WWI, but there are still around twenty working quarries extracting this stone. It is mainly used for paving. A number of the PLCM sandstones are massive and suitable as building stones in their own right, for example the Old Lawrence Rock quarried near Wigan in Lancashire (known as Appley Sandstone) and the Tom Bobbin Rock worked at Burnley. These sandstones are responsible for the regional character and appearance of the Lancashire mill towns and were widely used in the cities of Manchester and Liverpool. These building stones were used at in a relatively small scale for 19th and early 20th century construction in NW Wales, mainly for structures such as banks and other civic buildings.

Cefn Stone is a drab buff to bright yellow-coloured sandstone from the Bettisfield Formation of the PLCM which is quarried at Ruabon and at Bryn Teg Quarries at Broughton near Wrexham in Denbighshire. It is a medium- to fine-grained, sandstone, compositionally a lithic to sub-lithic arenite with feldspar, dark lithic fragments and sometimes speckled with orange iron oxides. Feldspar is generally weathered to kaolinite and sometimes weathered out and rare lithic mud clasts and plant fossils are occasionally encountered. It is more susceptible to weathering/pollution that Gwespyr Stone develops a blackened crust (Haycock, 2018; Howe, 1910). Cefn Stone has been used in the construction of the University of Bangor and in numerous Victorian churches in the region.

Pennant Sandstone, known as Blue Pennant or Gwrhyd Pennant, is quarried near Neath and is the Rhonda Member of the Pennant Sandstone Formation, itself part of the Warwickshire Group, the upper most sandstone group of the Carboniferous. This sandstone is a finely laminated blue-grey sandstone with well-developed ripples and related sedimentary structures. It is a sub-greywacke composed of quartz plus lithic clasts of schist and quartzite and

Figure 2.20 Riven slabs of Pennant Sandstone with ripples, used as paving on Twthil in Harlech.

abundant dispersed organic matter. Feldspar is absent. The stone is also variably iron stained. It is used as dimension stone and as flagstones. The latter, cut parallel to bedding, show off the sedimentary structures and are particularly attractive. Over the last decade, Blue Pennant has become increasingly popular for building and paving in North Wales (Figure 2.20).

2.11.2.3   The Cheshire sandstones

The Upper Palaeozoic to early Mesozoic Cheshire Basin contains a fill of Permian and Triassic clastic sediments of the Sherwood Sandstone Group and overlying Mercia Mudstone Group. These comprise a sequence of reddened sandstones, siltstones and mudstones accumulated in a half-graben tectonic setting. Much of the geographical area of the county of Chester is underlain by Permo-Triassic strata, and these lithologies have been widely quarried as vernacular building materials used locally and also exported outside the region, particularly into North Wales. The best building stones are derived from the Sherwood Sandstone Group. The base of this unit consists of a series of mottled red and cream-coloured, very soft sandstones, the lowermost Kinnerton Formation and the uppermost Wilmslow Formation. These units were only used very locally as building stones. The Kinnerton Formation is separated from the Wilmslow Formation by the Chester Pebble Beds Formation (Handley Stone). The Kinnerton, Chester and Wilmslow Formations belong to the traditional Bunter division of the Lower Triassic. The Chester Pebble Beds have been quarried since the Roman Period. This unit mainly outcrops in western Cheshire; stone was quarried in the city of Chester and also at Eccleston, Handley and Tattenhall, amongst many other locations. This is the main stone used to build both Roman Deva and modern Chester; a quarry in suburb of Handbridge on the banks of the River Dee is preserved and has a Roman relief carving of the goddess Minerva. It is a cross-bedded, red sandstone with strings of pebbles with up to 8 cm long axes. This characteristic allows the identification of this stone as the other red Cheshire sandstones are not conglomeratic, with the exception of the Delamere Member of Helsby Formation which, although superficially similar to the Chester Pebble Beds, does not contain pebbles larger than 2.5 cm (King, 2011).

Middle Triassic Keuper sandstones occupy the upper Sherwood Sandstone Group. These are the predominantly red, cross-bedded sandstones of the Helsby Formation. The Helsby Formation is the most widely quarried of the Cheshire sandstones, having widespread outcrop, particularly along the escarpments surrounding and within the Cheshire Basin. The lowermost member is the Passage Beds. These do not have continuous outcrop but are used where they occur as building stones. They are brown and pink, coarse-grained sandstones. Passage Beds were used in the construction of Beeston Castle (1220) which itself sits on an outcrop of this unit. The overlying Delamere and Frodsham Members vary from red-brown, pink or liver-coloured sandstones to buff or even grey sandstones in the upper part of the stratigraphy (Figure 2.21). Mottled red and cream sandstones also occur. This latter type was very popular for use in North Wales. It was used in the construction of Conwy Castle and also in several 19th and 20th century buildings in the towns of Conwy and Bangor. Delamere Member sandstones are red brown, with strings of small quartz pebbles (<2.5 cm). The Malpas Sandstone, a bright-red, cross-bedded sandstone is quarried from the Tarporley Formation, the lowest unit of the Keuper (middle Triassic) Mercia Mudstone Group.

When seen out of geological context and within an architectural context, the Cheshire sandstones can be difficult to differentiate from the red sandstones quarried from the Lower Carboniferous Basement Beds facies quarried at Moel-y-Don and Pwllfanogl

Figure 2.21 Red Cheshire sandstones with cross-bedding and contorted (slumped) bedding, used as dressings in the Royal Welsh Yacht Club, which is built into the town walls of Caernarfon.

on Anglesey and also the Gloddaeth Sandstones quarried on the Creuddyn Peninsula and used in Conwy and its environs. Greater familiarity amongst geologists with the Cheshire stones as opposed to the red beds of North Wales has led to many examples of local stones being identified as Cheshire sandstone and naturally to confusion and doubt as to the degree to which Cheshire sandstones was imported into North Wales.

2.11.3   English Jurassic limestones

Middle and upper Jurassic limestones have found much use as ashlar masonry, particularly for civic buildings and churches in England. The three main stones in this category are middle Jurassic Bath Stone and Cotswold Stones (Inferior Oolite Group) and the various facies of the Lincolnshire Limestone Formation as well as upper Jurassic Portland Stone (Portland Limestone Formation). These oolitic limestones were imported on a small scale to NW Wales but found relatively limited use. Bath Stone is used for window dressings in a few 19th and early 20th century churches throughout the region. Portland Stone (Upper Jurassic oolitic limestone from the

Isle of Portland, Dorset), though ubiquitous in civic building in most of England and in Cardiff, is found in very few buildings in North Wales. The former post office in Bangor is an exception, and it is used here as dressings (Figure 2.22) on a number of stone and brick-built buildings in Porthmadog. Painswick Stone, a variety of Cotswold stone and like Bath stone, a formation of the Inferior Oolite Group, is used in the interior of Penrhyn Castle. Lincolnshire Limestone is almost unknown in the region, despite the Earls of Ancaster having a seat in the region, at Gwydir Castle in the Conwy Valley. According to some sources, Ancaster Limestone is used at St Mary's Church Betws-y-Coed, but this is a misidentification.[8] The lack of this stone in NW Wales is probably more a consequence of the British transport network which is dominated by north-south trade routes. Transporting stone from east to west probably required more effort and expense than it was worth.

Fossiliferous, lithographic limestones from the Southern Frankonian Alb in Bavaria, Germany, generically known as Jura Marble are coming into use in the area, as in many places across Europe and beyond. This is an attractive and popular building stone

Figure 2.22 Portland stone is used for dressings in the old post office building in Bangor. The stone used here includes a fine specimen of the Upper Jurassic fossil bivalve Isognomon sp.

---

8 St Mary's Betws-y-Coed is constructed from local Hafod Las Stone with dressings and interior facings in mottled Kinnerton Sandstone from Cheshire. It also has a fine pulpit constructed from Delamere Sandstone dressed with Devon Marbles and a font of polished Cornish Serpentinite.

belonging to the Kimmeridgian Treuchtlingen Formation and has a rich fossil fauna of ammonites, nautiloids, belemnites and sponges, all of which are clearly observed in honed or polished surfaces. This stone has been used to clad the Pontio arts centre in Bangor.

2.11.4   English Alabaster

English Alabaster is a much forgotten ornamental stone, but it was a material of great value during the 14th–16th centuries and an important export commodity from England to the wider British Isles and continental Europe. Its main use at this time was for ecclesiastical carvings and specifically chest tombs memorials and wall plaques, often carved with effigies of the deceased. Stratigraphically, this stone is from the Tutbury Gypsum Beds of the Cropwell Bishop Formation. There is a long history of the quarrying of these stones from mines in the vicinity of Chellaston in Derbyshire and Fauld in Staffordshire in the English Midlands. English alabaster is used for funerary memorials in Llanpeblig Church in Caernarfon and St Mary's and St Nicholas's Church in Beaumaris.

2.11.5   Imported igneous and metamorphic rocks

By the later 19th century, igneous rocks from Scotland and Scandinavia were being used in the commercial districts of the main towns in NW Wales, where like the rest of the United Kingdom, they were very much in fashion. The invention of granite cutting and polishing techniques which took place in Scotland during the 19th century along with the expansion of the railway network brought these strong, hardwearing and decorative stones to all of the British Isles, where they were used, predominantly as decorative dressings and cladding as well as for gravestones, monuments, plaques and water fountains. Pink Caledonian Peterhead Granite from north Aberdeenshire is very commonly used as the attractive larvikite from the Oslo Graben region of Norway. The latter stone with its schillerescent feldspars was widely used from the 1890s onwards for cladding shop fronts and can be seen in Caernarfon, Bangor and Conwy. Swedish 'Coastal Red' Granites quarried on the Kalmar Coast to the south of Stockholm were also very popular with their brick-red colour imparted by potassic feldspars. They are a variety of rapakivi granite lacking ovoids, known as pyterlites which were intruded between 1.8 and 1.6 Ga into the Transscandinavian Igneous Belt. English Cornish Granite makes an occasional appearance as grave markers and also occasionally as ballast, along with Newry Granodiorite and the Wicklow granites from Ireland (see below).

At around the same time, decorative and coloured marbles were gaining in popularity in the decoration of interiors, particularly hotels, churches and town halls (see Watson, 1916). Although local Penmon, Halkyn and Mona 'marbles' remained popular locally, NW Wales did not entirely escape this trend, and imported stones were used throughout the region for, particularly, ecclesiastical fittings in the later 19th and early 20th centuries. Although outside the main area of study, the so-called 'Marble Church', St Margaret's at Bodelwyddan in Denbighshire, is a spectacular example of this fashionable use of decorative stone. Its exterior was built by John Gibson in 1860 from a porcellanous limestone from Llandulas Quarry (part of the Clwyd Limestone Group), and the interior includes over twenty varieties of decorative stone sourced from local deposits (i.e. Penmon marble), the wider British Isles and Europe (Shipton, 2019). The interior of Penrhyn Castle also makes good use of local and imported decorative stone, particularly used for fireplaces, decorative urns and table tops.

2.12 BALLAST

The importance of the global slate export trade from the ports of Porthmadog, Caernarfon, Port Dinorwic and Port Penrhyn and the general lack of imported goods to the region meant that the slate ships would arrive at port in ballast. Seawater or stone rubble was used to weigh the incoming slate ships. Ballast was often dumped off-shore before entering port and indeed an artificial island, Cei Ballast (also known as Lewis's Island), has been formed at Porthmadog. At Port Penrhyn, just to the north of Bangor, the slate quay juts out into the Menai Straits and the northern side of the ashlar-built harbour complex has been shored up with ballast. Similarly of the region of Port Dinorwic (Y Felinheli) directly north-east of the port is built of made ground reputedly constructed from ballast (Gwyn, 2015). Ballast is also used as a construction material in the port towns and villages. It is generally used in the construction of warehouses and walls in the harbour areas. The harbour buildings at Doc Fictoria in Caernarfon feature a range of one-off exotic stones which could only have been derived from ballast. A grey granitic rock is used in large quantities in Victorian domestic and civic buildings in Caernarfon, Port Dinorwic (Y Felinheli) and, to a lesser extent, in Bangor (Figure 2.23). The use of this stone was commented on by Greenly (1932) as being used in later 19th century restorations of Caernarfon Castle and also subsequently noted by Neaverson (1947), Nichol (2005) and Lott (2010). However, these authors did not investigate its use outside the castle, which is considerable. It is used for the construction of many of the 19th century administrative buildings in Caernarfon and for housing in Y Felinheli as well as being used for a number of churches and chapels in these towns. It is a medium-grained, pale grey granodiorite, composed of plagioclase, quartz and green hornblende, which is sometimes stained pink. Greenly (1932) reports that 'a certain John Jones' was employed to repair the Castle in the final quarter of the 19th century and procured his stone as cheaply as possible by buying ballast from incoming ships from Ireland. Analysis of thin sections of this stone strongly supports a provenance of the Newry Granodiorite quarried in County Down, Northern Ireland. Granitic rocks from the Wicklow Mountains of the Republic of Ireland also make an appearance, blue-grey in colour with their

Figure 2.23 Newry Granodiorite was bought to the slate ports of Caernarfon and Port Dinorwic in large quantities and is used as a building stone for walls as well as domestic and civic buildings.

distinctive, needle-like crystals of tourmaline and platy crystals of biotite. The expanding cities of Dublin and Belfast in the 18th and 19th centuries made much use of Welsh slate, and the rubble of their prime building stones was sent to Wales and used to build the slate ports. Slate from the Nantlle region in particular found one of its main markets in Dublin, shipped out from Caernarfon Slate Quay (Gwyn, 2015). There has not been a systematic geological or petrological study of the ballast found in the slate ports, but such a survey could potentially contribute much to the knowledge of the slate trade. One might also expect to find stone from the Baltic Regions and perhaps even North America and Australia.

Chapter 3

North West Wales before Edward I

3.1 INTRODUCTION

The landscapes of mainland North Wales and Anglesey are dotted with a large number of archaeological sites, dating from the Neolithic period onwards. These include funerary and possibly ritual monuments from the Neolithic, Bronze Age stone circles and Iron Age Hillforts. NW Wales was important as the western frontier of the Roman Empire and was well-garrisoned with a number of forts and other settlements. Towards the end of the Roman occupation of Wales and England, Christianity came to the region and was locally embraced, leaving a legacy of place names, churches and memorial stones. Many of these constructions or objects have endured *in situ* until the present day because they were constructed of, or carved from, stone. However, throughout this period, little knowledge of stone working and stone quarrying exists. We know that stone was quarried in various locations in the region during the Neolithic and Bronze Age for the production of lithic tools and that copper ore was worked during the Bronze Age and into the modern period on the Great Orme and at Parys on Anglesey. There are few other examples of ancient quarries and mines, however. Certainly the Roman builders and engineers understood stone quarrying methods and worked slate in the Llanberis area and limestone and sandstone from the Carboniferous strata of Anglesey, but any Roman workings have been obliterated by more recent stone working in the quarries. Much of the early stone, certainly that in the Prehistoric megalithic sites, was probably not quarried but collected locally as field stones, lying on the glaciated landscape. The lowland landscape we see now, much modified by the development

DOI: 10.1201/9781003002444-3

of agricultural land and the clearing and subsequent enclosure of fields by dry stone walls, gives an artificial picture of the frequency of field stones. However, the retreat of the glaciers 10,000 years ago would have left behind a landscape strewn with stone, from boulders to cobbles. Whether quarried *in situ* or brought to site by the action of the ice sheets, the complex and varied geology of NW Wales has provided communities living in this region over the last six millennia with a selection of unusual and uncommonly used building stones, some of which are unique to the region under study.

3.2 STONE BUILDING IN THE NEOLITHIC PERIOD

Palaeolithic peoples, including Neanderthals lived in North Wales as early as 230,000 BCE (230 ka), went on to occupy the landscape during interglacial periods. However, during cold periods, this region was fully glaciated, the summits of Snowdonia were completely buried by ice. Following the end of the last ice age (11 ka), Mesolithic hunter-gatherers were returning to Wales by as early as 7 ka, but their impact on the North Welsh landscape was minimal. Sites associated with Mesolithic lithic tools are predominantly coastal 'campsites' where beach deposits would have yielded a wide variety of lithologies suitable for tool-making, including cherty lithologies as well as shells. Nevertheless, studies of scatters and 'hoards' of lithic tools have revealed surprisingly scarce examples from North Wales during the Prehistoric period. Smith (2005) records a total of 340 lithic find spots. Much was to change with the dawn of the Neolithic. In Britain, the Neolithic period is defined from the early fourth millennium BCE[1] when people migrated from continental Europe, bringing with them sophisticated techniques for the manufacture of stone tools, farming, a more sedentary lifestyle and a developing tradition for building permanent funerary monuments from stone. An Atlantic-facing, 'megalithic culture' developed towards the end of the fifth millennium BCE in Western Iberia and/or the Brittany Peninsula of France. By the early fourth millennium BCE, these migrants had reached Ireland and NW Wales and have

---

1 Here we change our measurement of time from terminology used to measure geological time, measured in thousands and millions of years (ka and Ma) to the terminology used in archaeology and history, defined as time before the common era (BCE also BC), the year zero, to the Common Era (CE, also AD).

left behind them a significant number of funerary or ritual monuments constructed from 'megaliths', large stones used alone or in groups (Cunliffe, 2001).

In North Wales, megalithic sites date from c. 3,800 to 3,000 BCE, and monuments comprise stone circles, standing stones and burial chambers or dolmens; the latter types are primarily found on the island of Anglesey, but a number also exits on the western mainland and particularly on the Llyn Peninsula. These structures are known locally as cromlechs and comprise a (sub) horizontal capstone supported by three or more upright stones at heights of up to 1–2 m above the ground. These nowadays often freestanding structures were originally covered by a mound of earth or a cairn of rubble. Many were excavated in the early historical period in the hope of finding treasures. These tombs have been classified into various types based on the configuration and number of stones used and the shape of the resulting chamber. There are some twenty or so burial chambers on Anglesey including several recognised configurations of so-called 'long graves' and passage graves. Historical evidence suggests that there were many more burial chambers of which no trace survives, their stone almost certainly dressed and worked into walls and other structures (see Lynch, 1991). On the mainland there is a concentration of burial chambers on the Llyn Peninsula and the passage graves at Capel Garmon and Dyffryn Ardudwy which not only lie outside the area of study but outlie the main clusters of this type of monument in the region.

No systematic study of the lithologies used for the construction of megalithic monuments of Anglesey and the mainland has been made to date. However, there is little evidence to suggest that the often very large stones used were transported long distances for the purpose of construction. Outcrop in Anglesey particularly is sparse and largely covered in drift. However, the current agricultural landscape is deceptive and before farming, the region would have been littered with numerous field stones in the form of glacial erratics (which may not have travelled far from their place of origin) and frost-shattered outcrops. Large, slab-like stones could have been collected from the surface or levered and split from local schist outcrops. In the west of Anglesey, karstic weathering of the Clwyd Limestone Formation strata resulted in areas of limestone pavement, from which blocks could have been similarly collected. A number of monuments of this type fall within the area covered in this study, in clusters of sites close to the Menai Straits between

Brynsiencyn and Llandaniel Fab and in the vicinity of Moelfre on the north east coast of the island.

The passage grave at Bryn Celli Ddu was first formally excavated in 1925 by Wilfrid J. Hemp, the then secretary to the Royal Commission of Ancient and Historic Monuments in Wales (RCAHMW). The current turf-mound covered structure was built later by Hemp and is approximately half the diameter of the original mound structure (Figure 3.1). The central chamber with the dressed column of the 'Pillar Stone' would have stood close to the centre of the mound. The burial chamber is itself built upon a pre-existing henge, with a circle of stones and an encircling ditch. The interior also underwent some reconstruction during Hemp's programme of works. New excavations of the surrounding area have taken place (see Reynolds et al., 2016) and substantial evidence is coming to light of a ritual landscape in this local area, including a second henge structure to the south of the passage grave which, at the time of writing, is undergoing excavation.

During the 1920s, the geologist Edward Greenly visited the excavation at Hemp's request and identified the stones as locally derived (Upper Carboniferous) gritstone, 'hornblende schist' and glaucophane schist (Hemp, 1931). The slabs that form the interior, including Hemp (1931)'s 'spiral slab', incised with a small spiral glyph are made of the locally outcropping rocks of the high-pressure

Figure 3.1 Bryn Celli Ddu Burial Chamber, mound and henge. In the inner chamber and outlying stones are schists from the Penmynydd Terrane of Anglesey.

metamorphic blueschist-bearing Penmynydd Formation, and here at Bryn Celli Ddu include slabs of pure glaucophane schist in the stone fabric. The majority of the stone slabs used are composed of millimetre scale, alternating bands of glaucophane-rich and epidote-rich schist. This blue-and-green banding can be easily discerned by the naked eye on the Pillar Stone and other well-lit slabs within the grave structure.

At the western entrance, the so-called Pattern Stone, inscribed with zig-zag forms, stands separately from the main barrow. The modern 'Pattern Stone' standing at the edge of the mound is a replica in concrete; the original is now in the National Museum of Wales (NMW) collections.[2] The Pattern Stone was found inside the mound during excavations in the 1920s and archaeologists assumed that it was intended originally to be buried and therefore hidden (Hemp, 1931). At just over 1.5 m in height, it is a slab of gritty, red, iron-rich sandstone derived from the Upper Carboniferous 'Basement Beds' sandstones outcropping nearby on the shores of the Menai Straits. However, Edward Greenly indicated that it could have been derived from a 'quarry' in the field adjacent to Bryn Celli Ddu, which is located close to the unconformity between the Carboniferous and Neoproterozoic-Cambrian rocks.

A prominent, glaciated outcrop of schist is situated c. 150 m to the west of Bryn Celli Ddu, with an east-facing, sloped rock face formed by the dip of the metamorphic foliation. This surface is marked by incised cup-structures, which are probably contemporary with the burial mound. This slab of rock produces an echo of sharp sounds made close to the burial mound, and this, perhaps intentional, geoacoustic phenomenon has been explored by Devereux and Nash (2014). Recent excavations have shown that this outcrop was quarried and therefore substantially modified in the post-Medieval period (Reynolds et al., 2016).

A kilometre or so further west, two burial chambers are located on the estate of the Marquess of Anglesey at Plas Newydd: the large and spectacular double dolmen of Plas Newydd Burial Chamber and the Bryn-yr-Hen Bobl passage grave. Like Bryn Celli Ddu, the Plas Newydd Burial Chamber (Figure 3.2) is constructed of large slabs of blueschist. The facies here is a glaucophane-garnet schist.

2 Catalogue archNum 29.403.

Figure 3.2 Plas Newydd Burial Chamber is constructed from slabs of the locally outcropping blueschist facies rocks of the Penmynydd Terrane. Although locally abundant in this region of coastal Anglesey, blueschists are globally uncommon lithologies. There are few structures built from blueschist facies rocks, and the Neolithic funerary architecture of south Anglesey is an exceptional use of building materials within this region.

This group of two dolmens does not fit any recognised pattern of Neolithic construction or alignment, and it is entirely plausible that the structure was modified by 18th century landscape gardeners in order to make it a more attractive folly (Lynch, 1991).

Bryn yr Hen Bobl was excavated by Wilfrid Hemp between 1929 and 1934 (Hemp, 1935). Unlike the nearby Bryn Celli Ddu, Bryn yr Hen Bobl is located on Carboniferous limestone sub-crop, and this stone and cobbles derived from local boulder clays are the main construction materials used. The limestone is of a laminated nodular variety typical of the Loggerheads Limestone facies of Anglesey.

The nearby dolmen at Bodowyr is believed to have been a passage grave and the polygonal chamber and capstone survive. The stones are all a folded, felsic schist with foliation-parallel quartz veins. Again, this is derived locally from the schistose lithologies of the Berw Shear Zone and Aberffraw Terranes, which lie north west of and strike parallel to the Penmynydd Terrane. The same rock is used for the capstone of the Perthi Duon Dolmen which is located on sub-crop of Carboniferous limestone near Brynsiencyn

(Nash et al., 2014). This monument was once assumed to be an erratic but its designation as a dolmen has been proven by recent excavation.

Ty Newydd is another passage grave structure and has a particularly narrow capstone (probably only a fragment of the original) made of the Ordovician Treiorwerth Grits and is supported by upright stones of grit and one of pink, coarse-grained Coedana Granite. Further to the north, the two cromlechs at Presaddfed are located near the boundary of the Treiorwerth Grits and the unconformably underlying Gwna Greenschist Group, and they are constructed of these two lithologies. The capstone of the larger structure is composed of Treiorwerth Grits with supporting stones of chlorite schist.

In the area between Moelfre and Benllech, a style of dolmen architecture comprising a polygonal chamber constructed of short stones, surrounding an excavated pit and then covered by a large capstone seems to predominate, and examples of these can be seen in the Lligwy, Benllech and Glyn Cromlechs. All the stones are nodular limestone from the Loggerheads Limestone, and they show evidence of karstic weathering indicating that they were acquired from (very) local limestone pavement outcrop. The burial chamber at Lligwy has an enormous capstone, some 5 m across and almost a metre thick. The partly subterranean burial chamber is surrounded by ten relatively short uprights. The large capstone could not have been transported very far and was probably levered up from outcrop in the same field or indeed, jacked-up *in situ*. When excavated by E. Neil Baynes in 1909, the burial assemblage was found to be almost intact and the remains of up to thirty people were found, along with flint scrapers and pottery (Lynch, 1991).

On the mainland, Bachwen Portal Dolmen, near Clynnog-Fawr, is constructed on the coastal plain on an outcrop of Padarn Tuff Formation ignimbrites. These blocks would once again have been collected as field stones or levered out of coastal outcrops. Here the stone is a dark grey, foliated, vitric tuff with scattered clasts of quartz.

3.3 NEOLITHIC INDUSTRY: STONE TOOLS

In 1919, the geologist and antiquarian Samuel Hazzledine Warren (1873–1958) and his wife Agnes Mary (née Rainbow) were visiting Snowdonia on holiday. Whilst walking on Graig Lwyd near the Penmaenmawr Granodiorite quarries, Agnes found a fragment of

a Neolithic stone axe on the path, and Hazzledine Warren himself later found a partially finished axe surrounded by numerous
chipped flakes on the hillside nearby. He went on to find evidence
of stone-working debitage in 'nearly every molehill' and was the
first to discover the extensive and significant 'axe factory' now well
known in this area (Hazzledine Warren, 1919). The Penmaenmawr
Granodiorite body varies in grain size from a medium to coarse-
grained rock to a very fine-grained variety (200 μm), light blue-grey
in colour, it has a flinty appearance and breaks with a reasonable
conchoidal fracture. Although uniformly referred to in the archaeological literature as an 'augite granophyre', based on the petrology published by Hazzledine Warren (1919),[3] the rocks are better
described as quartz microdiorites, with plagioclase (andesine),
bastite pseudomorphs after enstatite, augite and quartz and potassic feldspar forming interstitial, granophyric intergrowths (Ball &
Merriman, 1989). The stone is speckled with sparse feldspar phenocrysts. The axes were being worked in this area during the late
Neolithic to Bronze Age and have been found locally in sites including a timber and earthwork henge structure at Parc Bryn Cegin
at Llandygai, where occupation has been dated to the mid to late
fourth millennium BCE (Williams et al., 2011) and at Bryn-yr-Hen
Bobl in the grounds of Plas Newydd and Bryn Celli Ddu on Anglesey (Lynch, 1991; Hemp, 1935). However, there is good evidence
that these stone axes were also widely exported, and they have been
found in contexts far from Penmaenmawr and Llandygai including
finds throughout Wales, in central, south and south-west England
and on the Isle of Man. Graig Lwyd stone tools have been assigned
Group VII in the classification of British stone axe lithologies (see
Clough & Cummins, 1988; Williams et al., 2011).

Another NW Welsh source of stone axes in the Neolithic is at
Mynydd Rhiw near Abersoch on the SW end of the Llyn Peninsula,
where quarrying pits follow the contact metamorphosed margins

3 Hazzledine Warren's petrological description of the Penmaenmawr axes reads
'sparsely distributed small phenocrysts of turbid plagioclase feldspar, small
rounded crystals and crystal groups of augite and still smaller decomposed
crystals of rhombic pyroxene in a microcrystalline matrix of quartz and feldspar with rods and isolated crystal of magnate'. 'Magnate' is a typographic
error and should read magnetite. Nevertheless, this antiquated petrological
description continues to be reproduced in literature on British stone axe heads
to the present day.

of a dyke. This is another important 'axe factory' and has been classified as Group XXI and utilised the Trygarn Formation hornfels
(Young et al., 2002). Olivine dolerite axes have been found in localities in Anglesey including at Bryn-yr-Hen Bobl (see J. G. Clarke in
Hemp, 1935) and are from local Tertiary dykes of this composition.

There are very few items of portable stone art known from the
North Welsh Neolithic. A singular exception is the Crochan Caffo
Stone, found in 1982 in the wall of a cottage garden. Believed to
date to the Late Neolithic to Early Bronze age, it is inscribed with
pecked, concentric circles (Nash, 2012). It is made of a fine-grained
sandstone of the Basement Beds of the variety typically found at
Malltraeth in Anglesey.

3.4   THE BRONZE AGE

In the second half of the third millennium BCE, a migration of
people from Western Europe brought new technology and a distinct style of ceramic 'beakers' into Britain. The so-called Beaker
People or Bell-Beaker People who arrived from the continent were
of a different physiognomy and brought with them new burial rites
as well as their distinctive pottery and were later to develop the
techniques of metal working (Parker-Pearson et al., 2016). Beakers
have been found in a number of the Neolithic burial chambers described above, indicating secondary use of these structures during
the Bronze Age. Several hundred Bronze Age burials and cremations have been found in the region under consideration here. These
generally contain single inhumations or cremations and associated grave goods. Burials are in round barrows, cairns and in cir-
cle-cairns and stone circles. Cremations are generally concentrated
into cemeteries (Tellier, 2018), and remains are placed in cinerary
urns. Ceramic petrology studies have conclusively shown that local
pottery production sites were in operation, due to the wares being
tempered with sands composed of grains of rock types including
glaucophane schists and olivine dolerites, lithologies distinctive of
and restricted to the unique geology of south west Anglesey (Williams & Jenkins, 1999).

With the exception of the stone circles and cairns, relatively
few stone-built structures exist from this period. Stone circles and
standing stones were built in the region from the Neolithic period
and throughout the Bronze Age, and some twelve or so examples

are extant within the broader region covered here, with three of these clustered around the axe factory at Penmaenmawr, including the large and almost complete so-called Druids' Circle, constructed of slabs of the local granodiorite. Burl (2000) recounts a number of destroyed circles on the Isle of Anglesey as well as solitary standing stones which were once part of larger groups of megaliths. Welsh stone circles are general elliptical rather than perfect circles and are associated with shallow banks and ditches and hence termed 'ring cairns'. Excavations have revealed inhumations as well as astronomical alignments of stones.

The spectacular Bryn Cader Faner is a combined rock cairn and stone circle (a 'circle-cairn') and dates to the later Bronze Age (Figure 3.3). It is located on the Cambrian strata of the Harlech Dome, a few kilometres north west of Harlech, on the moors above Talsarnau. Geologically it is on the dip-slope of a narrow ridge of Barmouth Formation quartz arenites, with the underlying Hafotty Formation forming the lower ground to the west and the overlying Gamlan Formation to the east. The Barmouth Formation outcrops just south of the circle cairn, dipping at a low angle towards the north-west. These stones, in angular chunks, are used to construct the cairn. They are very coarse-grained gritstones composed of poorly sorted, sub-angular quartz grains, with an overall grey colour, with some boulders cross-cut by thick bands of vein quartz.

Figure 3.3 The circle-cairn of Bryn Cader Faner, near Harlech. The cairn is constructed of upright slabs of Cambrian strata of the Harlech Dome.

The upright stones forming the circle are grey Hafotty Formation muddy siltstones, which have been predominantly split along their cleavage to form slabs.

Outside burial contexts, there is relatively little material culture associated with the Bronze Age. However, ubiquitous are Burnt Mounds, piles of heat shattered stones mixed with charcoal. These structures, incorporating locally gathered field stones, including cobbles from boulder clays, are obscure, and it is not clear what their function was, whether cooking sites (with hot stones used to heat water) or saunas, or whether they were associated with industry or indeed ritual practices.

3.5 BRONZE AGE INDUSTRY

Metal working came to the region in the Middle Bronze Age, probably from Ireland (Timberlake & Marshall, 2018), but despite the presence of substantial copper ore deposits in the region, little evidence of ore beneficiation and smelting has been discovered, and relatively few metal objects have been recovered from the region. As in many places, stone tools continued in use throughout the early and middle Bronze Age. It is quite possible that ore was worked seasonally and that it was traded by travelling copper miners. Copper is found at Parys Mountain on Anglesey and on the Great Orme, west of Llandudno, and both of these regions have been worked from the early Bronze Age well into the historical period.

On Anglesey, the ore deposits at Parys Mountain made this region the one of the main global copper producers of the 19th century, with ore processed and shipped out of Amlwch on the north coast of the island. The copper deposits are of the volcanic massive sulphide type (VMS), hosted in a series of Ordovician, Llandovery age, graptolite-bearing shales and pyroclastic rocks of a rhyolitic composition, which overly the greenschist facies rocks of the Mona Complex. Ores are either vein-hosted or occur as massive sulphide lenses and are composed of chalcopyrite, zinc sulphide and galena with quartz (Barrett et al., 2001). Evidence for Bronze Age workings is now well recognised, and hammer stones have been found in the northern area of the complex (Lynch, 1991). Evidence of fire-setting and Bronze Age drift working has also been identified. Bronze Age miners were probably working the products of supergene enrichment including native copper; such deposits have now been entirely worked out (Jenkins,

1995; Timberlake & Marshall, 2018; Barrett et al., 2001). Radiocarbon dates for Parys Mountain acquired by D. A Jenkins and quoted in Timberlake and Marshall (2017) suggest that copper exploitation may have begun as early as the 20th century BCE, making this region potentially the earliest site of metal production in the British Isles.

The copper deposits of the Great Orme near Llandudno are hosted in the Loggerheads Limestone (locally the Great Orme and Bishop's Quarry Limestone Formations) of the Clwyd Limestone Group. The mineralisation here is mainly epigenetic copper carbonate-rich ores, concentrated in a cluster of north-south trending veins rich in malachite, chalcopyrite and a number of accessory minerals, including pyrite, marcasite and ochres. Similarly, ore minerals infill vuggy cavities within the dolomitised limestones. Also important at the Great Orme is a black, earthy ore known locally as 'copper ddu' (black copper) which also occurs in veins cross-cutting the dolomites. Nodules of azurite are encountered in mudstone-rich beds within the local carbonate successions. Copper mineralisation superseded a Mississippi Valley-type lead-zinc mineralisation in the same area (Ixer, 2001). The resulting metals were copper with significant and distinctive traces of arsenic and nickel (Williams, 2018). Bronze Age miners could have extracted copper with ease from the malachite ores and possibly from the very soft and therefore easily extracted *copper ddu* deposits. Malachite, azurite and ochres would almost certainly have been used for pigments for painting and cosmetics too. Lewis (1996) reports that over 2,000 tools used by Bronze Age miners have been found in the earliest worked levels. These are predominantly large hammer stones, made from beach pebbles, variably described as 'hammers, mauls, pounders and crushers'. Bone tools and antler picks, stone lamps and ceramics were also found in excavations. Radiocarbon dates on bone collagen and wood charcoal have yielded calibrated dates indicating that mining on the Great Orme may have started as early as 1700 BCE and peaked between 1500 and 1300 BCE (Williams, 2018; Timberlake & Marshall, 2001). Recent work by Williams (2018) has shown that isotopic fingerprinting of Great Orme ores and metals indicates a distribution of bronze, chisel-headed axes ('palstaves') across central and southern Britain that closely mimics the distribution of Neolithic to Early Bronze Age stone axes from Graig Llwyd, a site located only 10 km north east of the Great Orme. This suggests that these two forms of tools of different materials were possibly traded together.

3.6 THE IRON AGE

In Britain, the Iron Age is subdivided into Early (~800–350 BCE), Middle (350–100 BCE) and Late stages (100 BCE–43 CE). It is defined as the time when iron production became prevalent through Europe and ended at the first invasion of Britain by Julius Caesar. However, in NW Wales, the division between Bronze and Iron Age was certainly transitional. During the Iron Age, we begin to see much more evidence of the settlement and management of the NW Welsh landscape, and indeed the earliest evidence for the construction of fortified structures, for which the region became famous in a later millennium. To paraphrase Aubrey Burl (Burl, 2000), the archaeological evidence in NW Wales seems to show that in the Neolithic and Bronze Age, people died but never lived, and in the Iron Age, people lived but never died. In contrast to the Neolithic and Bronze Age, the main evidence for occupation of the landscape is in the form of settlements of domestic hut circles, field kitchens and hillforts with little to no evidence of graves and other funerary or ritual practices. Evidence of iron production is not uncommon in the region, in the Late Iron Age and certainly into the Roman period, excavations have revealed evidence of small-scale smelting or, more commonly, smithing activity but no evidence of large-scale metal production (Crew, 1990).

Thousands of Iron Age 'hut circles' are located within the region, some enclosed in fortifications, others in small groups within an unfortified landscape. Roundhouses, typically around 7 m in diameter, could be constructed of wood, turf and clay and, of course, stone. Houses with stone-built foundations predominate in the Iron Age in many parts of North Wales. There has not been much in the way of systematic excavation of these settlements; Ghey et al. (2008) record around only fifty excavations of roundhouses in NW Wales. Associated material culture is sparse which has also made dating of these structures difficult, and some indeed may date back to the third millennium BC. Nevertheless, a peak in stone roundhouse construction occurs from 500 BCE to 500 CE, but stone roundhouse construction was always more prevalent during the NW Welsh Iron Age than it was in other parts of the United Kingdom (Ghey et al., 2008). These structures, generally much dilapidated when encountered in the field, are universally built from locally derived, rough, field stones. On the limestone sub-crop of Anglesey roundhouse foundations are sometimes built from upright slabs of the local limestone or mixtures of upright and flat-laid stone.

More substantial settlements dating to the pre-Roman Iron Age are the hillforts, of which there are some seventy examples in the area in question, and these are undoubtedly symbolic of local power structures (Ritchie, 2018; Smith, 2018). Hillforts are a common feature of the North Welsh archaeological landscape, and they have been built on lofty summits of the Llyn Peninsula and the low hills of the Arfon coastal plain and on Anglesey. All are located on ground higher than the local landscape. Some like Dinas Dinorwic sit on relatively low hills, only 170 m above sea-level, whereas the spectacular Tre'r Ceiri is located at 480 m, following a steep ascent and encloses 150 roundhouses. Promontory forts are located on coastal headlands where a defensive wall could be constructed across an isthmus. Hillforts may represent permanent settlements, some may have just been kraal-like structures for keeping livestock safe, whereas others may have been places to retreat to during times of conflict or danger from raiders. The latter interpretation stems from hillforts that are located away from good agricultural land. They are often long-lived features with archaeological evidence demonstrating occupation from the late Bronze Age into the Roman Period and maybe even beyond this into the Early Christian Period.

Typically hillforts enclose a relatively flat summit (although the interiors of some are rocky and rugged). They are clearly defensive sites with a combination of ramp-and-ditch earth-, timber- or stone-built encircling walls. As such they may be regarded as the first 'castles' of this region. Construction of perimeter walls in many cases produced earth ramparts, built from earth dug out from ditches and field stones. However, in some cases stone rampart construction may represent the earliest incidences of quarrying for local stone. This seems to be the case at Dinas Gynfor in north Anglesey, situated on a headland of somewhat unrewarding Ordovician shale and conglomerate beds. These were dug out to create a ditch and the stone then used for the rampart. Also on Anglesey, at Parciau Fort situated on limestone sub-crop, just south of Traeth Lligwy, the walls are constructed with upright slabs of limestone, again showing some evidence of quarrying (Lynch, 1991) and also resembling a palisade-type construction more typically built from timber. At forts such as Tre'r Ceiri and Conwy Mountain, both with walls enclosing villages of stone-foundation hut circles, the abundant field stones, formed from the frost shattering and glaciation of the igneous rocks forming these summits, were readily available. In both cases, Ordovician rhyolitic rocks were the main building stones.

Tre'r Ceiri above the village of Llanaelhearn on the eastern end of the Llyn Peninsula is an exceptional hillfort, in terms of its position on a high, steep hill but also in terms of its geology and use of stone. It is enclosed by substantial rubble masonry stone walls, an inner wall which is up to 4 m in height and at least a metre thick and a partial outer wall. The construction is from the slab-like rocks which were readily collected from the abundant scree deposits around the summit. An extraordinary stone-lined entrance passage cuts between the two walls, which is also paved with rubble. Hopewell (2018a) cites experimental archaeology conducted by the Gwynedd Archaeological Trust (GAT) which suggests that the wall, a circuit of 620 m circumference, could have been constructed in a period of 3 weeks by around one hundred dry-stone wall-builders. The summit of the hill of Tre'r Ceiri is composed of the felsitic Caergribin Rhyolite and Tre'r Ceiri Microgranite plutons; the former comprising the east-facing slopes and the latter the summit area and the west facing slopes. The slopes of the hill are covered in extensive scree deposits, and no quarrying of these hard and splintery rock types would have been required. The fine-grained, aphanitic Tre'r Ceiri Microgranite is the most abundant lithology used, primarily because it was available *in situ*, and it would have been easier to carry blocks down from the screes on the neighbouring hill of Yr Eifl than up from the screes on the east side of the Tre'r Ceiri summit. Another property of the Tre'r Ceiri Microgranite is the pronounced and large-scale conchoidal fracture which forms when this stone breaks; a feature which would have been of great use in the creation of natural water basins and quern stones (Figure 3.4).

Smaller forts, built on low-lying summits on the coastal plain, such as Dinas Dinorwig are constructed from earth manoeuvred from the thick blanket of glacial deposits. More remarkable from a Quaternary geological perspective is the fort of Dinas Dinlle which is situated on the coast a few miles south of Caernarfon. Currently half-lost to coastal erosion the site is built on and from a mound of glacial till, a push moraine, substantially thickened by glacial thrusting. These structures are well exposed in the beach cliffs (Harris et al., 1997).

Another ubiquitous feature of the Welsh Iron Age archaeological landscape is the occurrence of structures called 'Burnt Mounds'. These, as the name suggests, are mounds of stone, generally horseshoe-shaped, surrounding a small, flat area that was apparently a cooking hearth. The horseshoe commonly opens towards a small

Figure 3.4 The Iron Age Hillfort of Tre'r Ceiri. (a) Foundations of roundhouses constructed of the locally outcropping intrusive felsitic Tre'r Ceiri Microgranite. (b) This lithology has a pronounced conchoidal fracture which forms natural basins.

stream. Burnt mounds often seem to be located away from permanent settlements and are most readily interpreted as being field kitchens. These mounds are composed of locally derived, field stone rubble, which has clearly been subjected to heat and been burnt and is mixed with charcoal and blackened earth. Before iron cooling ware was common, heating food and liquids in containers out of direct contact with fire was an important technology. Fire-heated stones could be added to water to bring it to the boil or placed in pits with joints of meat to roast them. It is feasible that the mounds grew from discarded stones following cooking sessions. Despite the frequency of occurrence of burnt mound sites in the NW Welsh landscape, their dating is problematic, mainly due to their distance from settlements and the relative lack of associated material culture.

3.7   THE ROMAN OCCUPATION

The Romans arrived in North Wales in the second half of the 1st century CE, establishing over the next two centuries a network of roads and major forts at Caerhun near Conwy (Canovium), Caernarfon (Segontium) and at Tomen-y-Mur near Trawsfy-nydd. There were also numerous smaller forts on Anglesey and outside the area of study at Capel Curig in Snowdonia and Caer Gybi (Holyhead). As in other places, these forts would have been surrounded by a settlement or *vicus* (plural *vici*) with a largely non-military and probably non-Roman population. Vici would have varied in size depending on the size of the fort and the ser-vices that it required. Although a long way from the bright centre of the Roman Empire, NW Wales certainly embraced, to a certain degree, a sense of *Romanitas*, probably more so than is generally assumed. The Welsh word 'caer' signifies a Roman Fort or settle-ment, and this is encountered as a prefix in a significant number of toponyms. The Romans initially entered North Wales from Ches-ter (which is *Caer* in Welsh) and called by the Romans *Deva Vic-trix*. Deva was established in the CE 70s and was an active Roman centre until the Empire declined in the 5th century CE. Roman Deva was largely built from the local Triassic Chester Formation (Chester Pebble Beds), red, pebbly sandstones, and a quarry face at Handbridge on the River Dee has a relief shrine to the goddess Minerva, giving clear indication that local quarrying was active at this period. The position of the quarry on the River Dee also ensured easy shipping of stone from riverside wharfs. Here also early evidence of the use of slate, quarried from NE Wales, is seen used as a roofing material.

Prior to the Roman invasion, North Wales was occupied by people of the Iron Age Ordovices tribe. In 60 CE, an invasion force led by the general Gaius Suetonius Paulinus entered Wales with the aim of subduing the troublesome religious heartland of Anglesey.

This event was recorded in a few sentences by the Roman writer Tacitus, who relates that the Roman army were confronted by the Ordovices' troops, their priests ('Druids') and 'women, in black attire like the Furies, with hair dishevelled, waving brands' (transl. Jackson, 1937). Such an outlandish force momentarily paralysed the Romans with fear, but they rallied and retaliated and massacred the defending army and destroyed their sacred sites. Suetonius was then recalled to SE Britain to deal with the Iceni uprising in Colchester, and the Romans did not return to NW Wales and Anglesey until CE 77, led by the governor Agricola. This expedition had been triggered by the Ordovices rebelling and a subsequent massacre of the local cavalry garrisons. The retaliatory Roman force undertook to subdue the local troops. It was after this campaign that the fort at Segontium was built and a permanent Roman presence established in the region, which was to last for the next three centuries.

The main evidence for Roman occupation in NW Wales is in the form of forts or *castra*. These are of typically Roman, standardised design and layout; in map view their shape resembles a playing card, being rectangular with rounded corners. There would have been an earthwork bank, defining the perimeter of the fort which would initially have been walled with a wooden stockade which may subsequently have been replaced by stone. This is the *vallum*. A gateway, with watch towers, would be located in each of the four sides of the wall. Inside, buildings would be laid out in an orderly manner and linked by a grid network of streets. A *via principalis* or main street would normally occupy the long axis of the fort and link opposite principle gate houses. Inside the fort, the buildings would consist of a headquarters block the *principia* and the *praetorium* where the fort commander (*praetor*) lived and worked. Also nearby would be a *quaestorium*, the accommodation of the quartermaster or *quaestor*. Soldiers would have been housed in a series of barracks (or tents), and there would also have been grain stores (*horrea*) and other warehouses for storage of food and other supplies. Larger forts would have also contained latrines, canteens, religious buildings and even bath houses. Larger, permanent forts, with well-established *vici*, as at Chester, may well have evolved into as fully fledged Roman townships with a forum, villas, temples, amphitheatres and other civic buildings. The Roman Forts in NW Wales conform to this general layout, although extant remains are variable. At Canovium in the Conwy Valley, the earthwork *vallum* is clear, but here is little evidence now of interior buildings or a *vicus*. Far more substantial remains are found at Segontium in Caernarfon. At Tomen-y-Mur, in an isolated upland region north of Trawsfynydd, although there is little stonework remaining *in situ*, there are tantalising remains of a substantial military settlement complete with an amphitheatre.

3.7.1 Canovium

The fort of Canovium (also Kanovium) can be found at modern-day Caerhun on the banks of the River Conwy. The underlying bedrock is Silurian Denbigh Grits Formation greywackes and mudstones, but these are covered by a thick blanket of fluvial-glacial deposits. The original fort was a timber-built structure on earthwork embankments, established in CE 75. Stone construction of buildings within the perimeter of the fort began in the 2nd century CE. It commanded an important site both for trade and military purposes, situated on the banks of the river on an ancient river terrace. The River Conwy is notorious for its floods, and the fort is positioned high enough to avoid flooding. It is also just south of Tal-y-Cafn where the river narrows and shallows and ceases to become navigable for large craft. In addition to the regimental buildings of the fort, a substantial *vicus* was established in the vicinity, and a bath house has been excavated to the east of the fort (Baillie Reynolds, 1938). Very little associated stonework is visible at the site today. However, St Mary's Church and its small churchyard now occupy the NE corner of the fort and are possibly built from spolia from the fort buildings in the 14th–15th centuries (see Chapter 5). There is no obvious Roman masonry in the fabric of the church. The main building materials are dressed fluvial-glacial boulders composed of local volcanic and volcaniclastic rocks as well as Gloddaeth Purple Sandstone, an upper Carboniferous sandstone which could only have been quarried from Bodysgallen on the Creuddyn Peninsula.

3.7.2 Segontium

The Roman fort of Segontium in the modern town of Caernarfon was excavated by Mortimer Wheeler in 1923 and published the following year (Wheeler, 1924). This is by far the most important Roman site in the region in terms of both its architecture, its building materials and their legacy. Tradition has it that in the Middle Ages, Segontium was quarried for stone for the construction of Caernarfon

Castle and almost certainly for the neighbouring church at Llanbeblig. The reuse of stone in these buildings will be discussed in the following chapters. The fort is located on a low summit with view down the River Seiont and across the granite stock at Twthill, which had probably been occupied as a small fort or watchtower by the local people. Segontium is located on thick deposits of boulder clays overlying Nant Ffrancon Group siltstones and shales. The fort is very typical of its type, surrounded by a rectangular 'playing-card' earthwork *vallum*, with rounded corners. Arranged on a NE-SW long axis, gateways allowed entrances to the fort from all four sides. Evidence of occupation and building phases exists from the 1st–5th centuries CE. The bulk of the excavations lie to the north side of Constantine Road, whilst the remains of a bath house and a drainage system have been uncovered in the field to the south of the road.

In the late 1950s, a Mithraeum, now built over, was excavated by George Boon (Boon, 1960). This temple was associated with the main 3rd century building phase at Segontium but appears to go out of use in the 4th century, possibly with a general demise of the Segontium *vicus* and a transfer of the population across the Menai Straits to the trading post and *vicus* at Tai Cochion.

The area to the south of Constantine Road was excavated in the 1970s (Casey et al., 1993). A vicarage built in 1846 and subsequently demolished and its gardens had occupied this region during the time of Wheeler's excavations. Current excavations (see Kenney & Parry, 2012) are ongoing to the north and west of the fort, revealing more details of the *vicus,* work has progressed following the demolition of a school in this area.

Segontium was established in the Flavian period (69–96 CE) and probably founded c. 77 CE under the governorship of Gnaeus Iulius Agricola. It was large enough to support a garrison of 1,000 infantrymen at its peak occupation. Casey et al. (1993)'s excavations have done much to refine the stratigraphy, but many phases still remain obscure. However, it seems that the fort was abandoned soon after 393 CE (a date based on numismatic evidence from Roman forts across North Wales).

Today only building foundations remain (Figure 3.5) which remain uncovered only in the area north of Constantine Road; the excavations in the southern area were mostly backfilled (Figure 3.6). An exception is a cellar, the *sacellum* which remains largely intact below ground level and accessed by a stone stair. Nevertheless, the layout of the fort with a central headquarters building (*principia*),

Figure 3.5 The fort of Segontium in Caernarfon. For the most part, only foundation walls remain, nevertheless these provide a good ground plan of the fort and its buildings.

Figure 3.6 The sacellum at Segontium. This was a cellar probably used for storage. At the time of excavation, it was found to have been backfilled with rubble which has preserved its stonework. Note the coursed masonry of squared blocks of Basement Beds sandstones.

the Commandant's House (*praetorium*), and a series of buildings presumably used as barracks, workshops and granaries are clearly delineated. A room in the headquarters has been designated as a small shrine (*sacellum*) due to the discovery of an altar in the building's cellar (a later addition, created in the 3rd century CE). Also appended to this building in the 3rd century are the foundations of a solidly constructed hypocaust system. The function of the room overlying this structure is unknown, it may well have been a sauna or bath or a room heated for other purposes. In the southern section, the clear remains of a bath house are exposed. The main findings of the archaeological excavations in the 1970s were an architecturally impressive building with a central courtyard, immediately to the north of the bath house and dating to the late 3rd century.

The exposed foundations of the fort of Segontium are constructed of concrete-bonded rubble masonry (*opus incertum*), and Wheeler was able to subdivide the architectural styles of the walls into three groups. Group 1 walls are ascribed to the earliest stage of stone building in the 2nd century CE. Local stones and a red sandstone are the main building stones recorded in these structures which are best seen in the foundations of the Commandant's House. Characteristically, these walls are quite narrow, at around 60 cm width. Group 2 walls have been ascribed a 3rd century date and are comprised primarily of local sandstones and limestones. These structures are considerably more substantial to those of the earliest phase and are 75–80 cm in width. Buildings associated with this building phase appear to have slate, rather than tile, roofs. Group 3 walls are associate with a building phase in the late 4th century and characteristically use *spolia*, reused stones from older buildings, and therefore have a somewhat chaotic appearance with a jumble of stone rubble. All the walls are constructed using the standard Roman *opus incertum* technique, with a rubble masonry and lime cement core and external facing stones. It should be noted that Casey et al. (1993) have identified eleven phases of construction, however from a building materials standpoint, Wheeler's simplistic stratigraphy remains useful.

The use of slate as a building material was recorded by Mortimer Wheeler from the 2nd century CE in structures associated with Group 2 walls. It was used for flooring in the *sacellum* and also for lining the flue walls in the hypocaust building, but it was also used as roof tiles with either hexagonal, pentagonal or oblong form, classic shapes for Roman architecture (see Casey et al., 1993).

Roofing slates were subsequently found recycled into 4th century structures and at the excavations of the Mithraeum by Boon (1960). Purple slates from the Llanberis Slate Formation were used, but of a type sandier and less fissile in nature than those worked from Dinorwig in the 19th century. Edward Greenly, on examining the material excavated by Wheeler, suggested that they came from beds lower in the sequence than the slate typically quarried from the 18th century onwards. The lower units, though poorer-quality slate form more prominent outcrops and would have been more easily observed in the landscape. These slates could have been derived from Nant Peris or from Nantlle (Gwyn, 2015). These finds indicate some of the earliest quarrying activity within the Llanberis Slate Formation.

A fragment of a slab bearing a somewhat primitive relief carving of a figure bearing a spear and a plumed helmet, identified as the god Mars was discovered in the 1970s excavations of the southern area of Segontium (Casey et al., 1993). This object, assigned to the 2nd–3rd centuries CE, was possibly a gravestone and is carved from a grey, silty mudstone, which was also probably derived from the Llanberis Slate Formation.

Limestones from the Clwyd Group are one of the two main groups of building stone used at Segontium, presumably quarried at Penmon and brought by ship along the Menai Straits to wharves at Caernarfon. Blocks of Penmon limestone often with characteristic fossils occur frequently in walls of all periods but particularly in Group 2 and 3 walls. More flaggy facies of the Clwyd Limestone Group are also used for paving and examples still exist *in situ* around the well in the *Praetorium* courtyard. These come from the Leete Limestone and are particularly rich in the large fossil brachiopod *Daviesiella llangollensis.*

Along with limestone, gritty sandstones, conglomerates and breccias from the underlying and interbedded Basement Beds are used throughout the extant foundations of the fort as rubble masonry, and presumably as ashlar masonry, now removed. This stone was also used for column bases in the *praetorium* and *principia* as well as at the Mithraeum (Figure 3.7). There are a number of these kept today in the small lapidarium located in front of the Segontium Museum. These contain square slots which would have supported wooden posts and would have formed the foundations for arcaded verandas and porches to the principal buildings of the fort.

Portable stone objects have also been retrieved from excavations at Segontium. Six stone altars have been found in the main

Figure 3.7 Column bases excavated from the administrative buildings at Segontium and now collected in the lapidarium
outside the museum building. These blocks are cut from
Basement Beds gritstones and have a square basal section.
They would have supported wooden posts for supporting
a veranda surrounding the building.

site and at the Mithraeum with another one discovered within the
fabric of Llanbeblig Church. The best preserved is the Segontium
altar, which was found, not *in situ*, amongst rubble in the *sacellum* fill. A just over 40 cm tall, it has a dedication to the goddess
Minerva from one Aurelius Sabinianus, 'actarius', possibly one of
the fort quartermasters (Wheeler, 1924). There are traces of white
plaster on the altar and also traces of red paint in the incised lettering. It is composed of a medium to coarse-grained, well-sorted
sandstone with scarce but prominent protrusions of sub-rounded,
quartz pebbles up to 10 mm diameter and evidence of a white clay
in the matrix. It appears to be typical of the sandstone beds worked
from the Lower Carboniferous Basement Beds sandstone facies
outcropping at Malltraeth in Anglesey. Four fragmentary sandstone altars were also found at the Mithraeum and a fifth, 60 cm
tall, was found in the footings of Llanbeblig Church. These were all
constructed of identical sandstone to the Segontium Minerva altar,
described by the excavator as 'a local grit' (Boon, 1960).

Three fragments of rotary querns as well as some other blocks
that may represent reworked quern stones were excavated by Casey

et al. (1993). These were all made from local Basement Beds coarse-grained gritty and quartz-pebble rich, conglomeratic sandstones.

Fragments of red sandstones are scattered throughout the rubble masonry but occur primarily in the Commandant's House. Remaining examples show scant evidence of dressing or shaping. Mortimer Wheeler believed that this stone was brought from the Triassic strata of Chester. More plausibly it could have been extracted from the now quarried-out Lower Carboniferous Sandstones of the Basement Beds at Moel-y-Don. Arguably stones from both sources occur here. In his attempts to provenance stone used for the construction of Segontium, he invited Edward Greenly and Sir Aubrey Strahan to visit the site and study the building stones. Strahan, clearly somewhat stumped, rather tentatively identified the red sandstone used at Segontium as the Middle Triassic Delamere Sandstone Member from Cheshire (Wheeler, 1924). Strahan would have been less familiar with the then poorly described Basement Beds and their variation in facies, something that Greenly himself was later to further research and publish. However, the author has not seen any evidence of the distinctive red Cheshire Sandstone in the remains of the fort. Nevertheless, it is used in small amounts for early renovations to Caernarfon Castle which led Greenly to believe that it was *spolia* from the Roman Fort, though contemporary records for the construction of the castle dispute this (see Chapter 2). It seems more likely that the red sandstone used at Segontium is from the local quarries of the Basement Beds at Moel-y-Don which were subsequently worked out and would therefore have been stones unfamiliar to geologists in the 1920s.

Given their local abundance in the boulder clays covering the local area and also their suitability for *opus incertum* construction, relatively few glacial cobbles and boulders are found on the site. A few do occur in the wall cores of Wheelers Groups 1 and 2 walls and in both the cores and facing stones of Group 3 walls. These boulders are sub-rounded and are ultimately derived from the Ordovician volcanic successions of the immediate hinterland of Snowdonia. A few blocks of Twthill Granite are also found in wall cores of 3rd century or later walls; however, there is no evidence for active quarrying of Twthill Granite at this time, and so it is assumed that this was acquired as field stone. A single and rather large (70 cm long) slab of roughly hewn greenschist now lies on the ruins of the North West Gate. This stone is typical of that found within the Mona Complex of central Anglesey, and therefore it is unlikely that it would have been deposited east of its outcrop by the west-moving Pleistocene ice-sheets. It is possible that this block of metamorphic rock was intentionally brought to the fort of Segontium and perhaps used as a lintel in the gate. One might speculate that it originally came from one of the megalithic structures of SW Anglesey.

A regionally unique and unexpected use of tufa as a building stone occurs at Segontium and forms a limited and irregular stringer course in the 3rd century CE *sacellum* cellar (Figure 3.8). It is quite possible that these stones have been reused from an earlier building. Locally, tufa forms in hard water springs dominated by bryophytes; the moss *Cratoneurion commutati* in petrifying springs, and these are known in the mountainous areas to the east of Caernarfon. However, these do not arise from calcareous rocks and do not produce tufas in quarriable amounts (Farr et al., 2014; Gareth Farr, *Personal Communication*). Elsewhere in North Wales, extensive tufa deposits are known from Caerwys in Flintshire (Brasier et al., 2015) and also small deposits occur at White Beach (Traeth Fedw Fawr) on Anglesey near Penmon (Graham & Farr, 2014). Given the small amount of tufa used in the Segontium *sacellum*, it is impossible to locate the source, but assuming the Penmon region was quarried for limestone ashlars and paving flags, tufas from this area are the most likely candidates for the origin of this stone.

Figure 3.8 Tufa used in the construction of the sacellum at Segontium.

3.7.3   Tomen-y-Mur

The earth-built Roman fort at Tomen-y-Mur near Ffestiniog dates from 79 CE, and it was occupied until the late 2nd century CE. The site was also occupied in the Norman Period when a motte and bailey wooden castle were constructed on the site. The fort is located in the northernmost part of the Harlech Dome succession, situated on the boundary between the Gamlan, Rhinog and Maentwrog Formations. The Gamlan Formation is composed of grey-green to purplish, pyrite bearing mudstones with a weak slaty cleavage, interbedded by coarse, gritty quartz arenites. These formations of the Harlech Grits Group are in faulted contact with the Maentwrog Formation of the Mawddach Group, via the Trawsfynydd Fault Zone. These are the so-called 'Lingula Flags', silty, grey mudstones, with a few quartzo-feldspathic sandstone beds. There are also a number of small, sheet-like dolerite and microdiorite intrusions in the vicinity (Howells & Smith, 1997). Tomen-y- Mur's main phase of excavation took place in the 1870s under the general direction of Reverend E. L. Barnwell and then later, in diggings conducted by a 'Mr Holland or a Dr Lloyd' (Gresham, 1938). Unfortunately, these excavations were not recorded in detail, and many associated finds have now been lost. Nevertheless, the main structure of the Roman-period fortifications and an amphitheatre, the only example of its kind known to be associated with a fort in Roman Britain, and a local road network were mapped. Gresham (1938) provides a review of work undertaken on Tomen-y-Mur as well as recording an early survey of the site undertaken by members of the Cambrian Archaeological Society (Gresham, 1938).

The Tomen-y-Mur was constructed in two phases. An early structure was built in wood with turf and earthwork banks. The fort was rebuilt in stone in the early second century to illustrate CE, and pottery found in the vicinity secures this date. However, there are now no visible remains of Roman masonry *in situ* on the site. A wall to illustrate the form of the original *vallum* has been recently reconstructed in Rhinog Formation 'Harlech Grits'. Original stonework was robbed out for construction of local walls and presumably for building the motte of the Norman castle. The amphitheatre located, 300 m to the north west of the fort, was a timber-built structure erected on earthwork banks.

Gresham (1938) tentatively locates a quarry site within a band the Gamlan Formation gritstones in the footwall wall of the

Trawsfynydd Fault and whether from here or not, good building stone was certainly locally available. A large number of roof tiles were found on the site, and archaeologists had located Roman-period tile kilns at Pen-y-Stryd, 4 km distant. Ceramic tiles could have been used for paving or roofing. However, it is likely that the local Cambrian sandstones and grits were employed for building works, derived from the Rhinog and Gamlan Formations. Gresham (1938) notes that nine inscribed stones were retrieved from the site. He writes 'Six of these are well-known, having been walled-up in Harlech Castle. The other three have apparently not been seen for a great many years. They are built into the foot of the terrace at Plas Tan-y-Bwlch [a local farmhouse], in the first or most south-westerly bay and are now covered by the thick stems and leaves of a large magnolia'. The stones used in the walls of Harlech Castle, where it seems they were taken for display purposes in the mid-19th century, have subsequently been removed to the National Museum of Wales, as have those subsequently retrieved from Plas Tan-y-Bwlch (Hassall & Tomlin, 1977).

3.7.4   Small Roman forts and Romano-British settlements

Excavations taking place at the time of writing are starting to reveal the *vicus* surrounding the fort at Segontium (Kenney & Parry, 2012), and recent (limited) excavations and more extensive geophysical surveys on the Anglesey shore of the Menai Straits at Tai Cochion (opposite Caernarfon) have revealed a surprisingly extensive and well-established trading settlement (Hopewell, 2018b). Such *vici* are the only types of settlement associated with the Roman period in North Wales and the village at Tai Cochion would probably have been a regional centre in the 2nd–4th centuries CE. Only earthworks remain at the small garrison fort at Caer Leb, located to the north east of Tai Cochion. A bath house, presumably associated with a villa or even an inn (*mansio*), was excavated near Tremadoc in the early 1900s, and in 2012 a stack of typical, diamond-shaped Roman-period roofing slates from the Cambrian Llanberis Slate Formation, were discovered during the construction of the Porthmadog bypass, which were likely to have been associated with this site (Parry & Kenney, 2012; Chapman et al., 2013). They have been identified as coming from a bed in the slates known as the Blue Vein and were probably quarried from the Nantlle area (Gwyn, 2015).

The presence of Roman coins found on a number of Iron Age Hillforts in the region suggests that the occupation of these structures continued well into the Roman period and possibly after. The same could be said of roundhouses such as those at Penmon excavated in the early 1930s (Phillips, 1932, see below). The most important example of local domestic architecture in the region is the settlement of Din Lligwy on the north coast of Anglesey, located on the limestone sub-crop just inland from the modern quarry sites on the coast at Moelfre (Figure 3.9). Like the nearby Lligwy burial chamber, this settlement is constructed from the Loggerheads Limestone, much of the building material could probably then have been extracted from karstic surface exposure. Din Lligwy, excavated in 1905 by Neil Baynes (Baynes, 1908), and subsequently partially reconstructed for public access and interpretation, consists of two roundhouses and up to seven rectangular buildings within a pentagonal enclosure wall. The roundhouses are constructed from upright and horizontally laid slabs of Loggerheads Limestone in a similar manner to those used at the nearby hillforts at Din Silwy and Parciau. The rectilinear buildings are similarly built, with foundations of field stones with some uprights at doorways.

Figure 3.9 A roundhouse in the Romano-British settlement of Din Lligwy. Note the upright slabs of Loggerheads Limestone which form the foundations.

The settlement is situated on the edge of the Loggerheads Limestone which here forms a limestone cliff on the south-east bank of
the valley of the River Lligwy (Afon Lligwy), overlying the gritstones and conglomerates of the Basement Beds and a limited outcrop of Devonian Old Red Sandstone (the latter is poorly exposed
the river valley below). Arguably this site could be considered as
belonging to the Roman period or perhaps more accurately of
Romano-British occupation; the majority of the finds are of Romano-British and Roman pottery as well as a few Roman coins. Iron
ore and slag have also been found on site, suggesting that at least
one of the buildings was used for smithing (Crew, 1990). Numismatic evidence as well as pottery indicates settlement late into the
3rd or 4th centuries CE. However, architecturally the roundhouses
suggest occupation by local people with the addition of Roman-
ised, rectangular workshops. However, it would be a stretch of the
imagination to envisage the hand of a Roman builder at work here.

A number of Iron Age hut circles were excavated above and west
of Penmon Priory in the area known as Parc Dinmor in the 1930s
which were dated to the Romano-British period (Phillips, 1932).
Phillips's excavation was a rescue excavation as the archaeological
site was in the path of the encroaching quarry face, during a period
of high demand for Penmon limestone, required for the embank-
ment of the River Mersey in Liverpool. The excavation revealed a
hut circle which had a floor of the limestone bedrock (Cefn Mawr
Limestone). The foundations of the huts themselves were constructed
of this underlying limestone, which had been levered out along orthogonal joints and beddings planes. Indeed, Phillips found a small
quarry which could have supplied the stone but had subsequently
been used as a rubbish dump for discarded sea shells, shellfish being once the main food supply of the occupants of the village. As
further evidence of food preparation, thousands of limestone beach
pebbles were found around the site, brought up as pot-boiler stones.[4]
A saddle quern made from Basement Beds gritstone was also found
as well as a number of hammer stones, and slabs of red sandstone
(undifferentiated) that had cut marks, indicating they were chopping
boards. A Roman-style iron sickle was found (identified by Mortimer Wheeler), but no other traces of *Romanitas* were found. In fact,

---

4 Stones were heated in a fire until very hot and then placed in water to bring it to
the boil. This was a useful way of heating water in wooden vessels which could
not be placed in direct contact with fire. Pot boiler stones show evidence of heat
shattering and spoliation on their surfaces.

a complete absence of pottery was found on the site making dating difficult, but one may speculate that the village post-dated the Romans and may have been used into 5th or even 6th centuries.

Despite their smallness and seeming insignificance, these sites indicate a number of scattered homesteads and farms across the Conwy Valley, Arfon and Eifionydd where an intrinsic sense of *'Romanitas'* was adopted. During this period of Roman occupation, for the first time, we see evidence (albeit it scant) of stone imported into the Arfon region from Cheshire and clear exploitation of the local slate deposits.

3.8 IRON AGE AND ROMAN INDUSTRY

Frustratingly there is little archaeological or documented evidence for Roman Quarrying or mining in NW Wales. Nevertheless, it cannot be denied that the Romans came to the British Isles in the 1st century BCE with a good understanding of quarrying, stone-working or mining, and across Roman archaeological sites in Britain careful exploitation of local stone is evident. It is almost certain that the Romans quarried Basement Beds sandstones and grits as well as limestones, probably from the Penmon area of NE Anglesey. However, evidence of this is not clear and largely obliterated by more recent quarrying activity, and there is apparently little evidence of Roman-period tools or quarry marks in quarries. The area of St Seiriol's Well at Penmon (see below) is surrounded by a vertical rock wall of Cefn Mawr Limestone that is clearly a quarry face rather than a natural rock formation, although it has been sufficiently weathered and does not reveal any tool marks. It is possible that this dates to the Roman period. Similarly, it is also possible that the Romans mined copper from the Great Orme, although again direct evidence for this is lacking. Certainly, copper was extracted from Parys Mountain as copper cakes with Roman stamps have been found in the vicinity (Barrett et al., 2001). Evidence for Romano-British iron working has been found at Din Lligwy on Anglesey, where the local limestone could have proved useful as a flux.

3.9 THE EARLY CHRISTIAN TO NORMAN PERIODS

Christianity came to NW Wales, as it did to the rest of the British Isles during the Roman occupation and remained after the Roman Empire declined. Unlike much of mainland Britain, there was

apparently continuity in religious activity in Wales and the establishment of an early church and traditions of monasticism began in
the 4th and 5th centuries which was coincident with the development of the *llan*, a settlement developed by a community of Chris-
tians on sanctified ground. Characteristically, *llannau* surrounded
a church or similar holy site, frequently a sacred well or spring or
an area blessed by a figure regarded as saintly. Subsequently Llan-
occurs as a prefix to so many Welsh place names. Early *llanau* were
constructed from earthwork embankments and some were perhaps
fortified by a stockade. This landform can still occasionally be
discerned in the landscape today. A superb example of the preservation of such a structure is the site of St Cristiolus's Church at
Llangristiolus and its graveyard (see below).

Politically, a distinct Kingdom of Gwynedd had developed by
the 6th century; King Maelgwn was reckoned amongst the great
leaders on the island of Britain (Avent et al., 2011), and his descend-
ants ruled the area until the invasion of Edward I in the last quarter
of the 13th century. By the 'Age of the Saints' in the 6th–7th centuries, the Celtic Church and a tradition of monasticism were well
established, with important mother-churches founded within the
region under study at Clynnog-Fawr, Bangor and Penmon. Much
work was done here by early leaders of the Church, men who became important saints in the region, such as Beuno, Deiniol and
Seiriol, who have given their names to a number of localities in the
region. There were many, many other saints; Bardsey Island located
of the western end of the Llyn Peninsula is said to be the burial
place of 20,000 saints and became an important site of pilgrimage.
The pilgrims' way started at St Beuno's Church (and burial place)
in Clynnog-Fawr in the eastern Llyn Peninsula. Sainthood in Wales
was attached to men and women who were leaders in the church
and regarded to have led a particularly aesthetic life, rather than the
continental Catholic tradition of individuals who had met a grisly
end directly relating to their persecution as believers in Christianity. As the Saxons brought a new language to eastern Britain, the
Welsh language, once more common across Britain, became associated with the region, and this period very probably holds the
origins of the Welsh literature tradition. The border of Wales with
the Kingdom of Mercia and the rest of what would eventually become England was defined in the 8th century by an earthwork dyke,
constructed by Offa, King of Mercia (d. 797). Although built by a
Mercian King to keep the Welsh in, this immense structure enabled

clear boundaries to be placed around the country of Wales. A history of this obscure period has been pieced together from diverse and dispersed records by Davies (1993). Despite the lack of written evidence, material evidence suggests that Gwynedd was during this time well connected with the rest of Europe; Latin also remained in use and clearly, therefore, some members of the local communities were literate and in contact with the wider British Isles.

Inscribed and incised stones are the main artefacts associated with this early Christian period (5th–10th centuries). In North Wales, the great majority of these stones (some 150 examples are known) show carved patterns (often crosses) and lettering and many have Latin inscriptions. A small number of these stones are bilingual, inscribed in both Latin and in the early medieval Celtic alphabet, Ogham. These objects are of importance as they demonstrate a literate society and also a continuity in the knowledge of the use of stone from a period where little else survives. The corpus of stones has been published for the region by Edwards (2013) with a study and discussion of their geology by Horák (2013). The purpose of these stones is either as grave markers or memorial stelae, and the great majority are made from stones outcropping very locally to where they were found. Nevertheless, this includes some interesting lithologies and innovative use of local stone.

Horák (2013) has made a detailed study of these monuments in terms of their geology and has demonstrated a clear change in the use of stone materials over time. In the 5th–9th centuries, a wide range of stones are represented in the construction of the 59 monuments studied. These are in the majority sandstones and siltstones from a wide range of sources; stones found on Anglesey and Arfon are constructed from Ordovician and Lower Carboniferous sandstones, but with a significant proportion of stones, perhaps surprisingly, being constructed from much harder igneous rocks. Of the forty-two known stones dating from the 5th to 7th centuries, 40% are made from igneous rocks. From the 9th century, the gritstones and sandstones of the Lower Carboniferous Basement Beds are the dominant materials used for the manufacture of these stones. Between the 7th and 12th centuries, three quarters of the fifty or so known stones are made from these clastic sediments, with the remainder made from Lower Palaeozoic sandstones and siltstones. Surprisingly few of these stones are carved on limestone slabs, despite the much higher degree of workability of this lithology and the relative ease of letter-cutting on limestone surfaces.

At St Baglan's Church near Caernarfon, the fabric of which dates to the 16th and 17th centuries (see Chapter 5), a late 5th century grave marker commemorating 'Anatemorus, son of Lovernius' has been cemented into the wall and forms the interior lintel of the north doorway. It is made from a dark green porphyry which weathers to a cream colour. Two stones at Llanaelhearn Church (see below) are made from pink-weathering, grey porphyry of similar appearance. These stones were quarried locally from the Nefyn Cluster intrusions or perhaps associated crystal-lithic tuffs. Petrologically the 'Melitu' stone at Llanaelhearn resembles the Gyrn Ddu Porphyry, a stone also used for the 10th century sundial now located at St Beuno's Church at Clynnog Fawr (see Chapter 5). Also at St Baglan's is a 7th–9th century inscribed cross (with additional ship graffiti) which is located in the porch and carved from a fine to medium-grained sandstone from the Basement Beds, probably sourced relatively locally, on the Anglesey coast of the Menai Straits.

Although slightly outside the area of interest here, on the central Llyn Peninsula, the Llanor Stones are of great urban geological interest. Two stelae, known as the Llanor (or Penprys) stones, are now housed in the Victorian mansion and art gallery at Plas Glyn-y-Weddw in Llanbedrog on the Llyn Peninsula. These were found buried in a field at Llanor 6 km to the north-west in 1833. Between then and arriving at Plas Glyn-y-Weddw in 1993, the stones were reinterred several times and spent some time on display at the University of Oxford's Ashmolean Museum. The stones are 1.7 and 1.56 m tall respectively. The tallest one is composed of a rhyolitic tuff of trachydacitic composition. It is described with one word, 'Vedesetli', probably the name of a saint. The second stone is of similar composition though geochemically more silica-rich and rhyolitic in composition. It carries the inscription '*Iovenali fili, Eterni hic iacet*, which translates as 'Here lies Iovenalis son of Eternus' (Edwards, 2013). What is remarkable about these two stelae is that they are both columnar joints, whose shape has not been further modified, extracted as ready-made columns from pyroclastic rocks, ignimbrites, derived from the Ordovician Llanbedrog Volcanic Group and probably specifically from the Carneddol Rhyolitic Tuff Formation (Horák, 2013). The flat surfaces of the polygonal columns have provided natural smooth and flat surfaces for the letter-cutter to work upon. These stone have been dated to the late 5th–6th centuries.

Beuno was an important founding father of Early Christian Wales, and his burial place is reputed to be at Clynnog-Fawr, now the site of an impressive 15th century Church (see Chapter 5). Beuno was sanctified after his death. At Clynnog, there is no evidence of an Early Christian *llan*, but excavations in the early 20th century revealed foundations beneath the Chapel which may date to this period. The site is associated with a number of incised stones dating to this period. Inside the Capel-y-Bedd is Maen Beuno (Beuno's Stone, also known as the Waunfawr Stone; Edwards, 2013) which bears an inscribed cross. This stone is much travelled; it was bought to the church from Aberglaslyn Hall in Beddgelert in 1919. Prior to that it had been in Bodwyn hall and on the banks of stream at Glan Beuno, both localities in Bontnewydd near Caernarfon. It is a pale-brown weathering, well-rounded glacial erratic, ~100 cm high and 65 cm wide (Edwards, 2013). Unfortunately the continual laying on of hands over the centuries means that it has developed too much of a patina for the stone to be identified, but it is likely to be a volcanic or volcaniclastic rock from the hinterland of Snowdonia. Another large glacial boulder, 1.2 m in length, of volcaniclastic rock, with a white-weathering crust, is located outside the doorway in the north transept of St Beuno's Church, and this probably dates to the 8th century. It was found at the crossing of the church buried so that the upper surface of the stone was at floor level. Its significance in this location is unknown (RCAHMW, 1960).

In the churchyard at St Beuno's in Clynnog-Fawr, a spectacular sundial, said to date from the 10th to 12th century, is located at the SW corner of the Capel-y-Bedd, sadly sited in a spot heavily shaded by trees (Figure 3.10). The dial (which is at least south facing) is incised onto a squared slab of microgranite porphyry quarried locally at Gyrn Ddu. The size and shape of the rectangular stone slab, standing almost 2 m in height above ground, 50 cm wide and with a uniform thickness of 12 cm, show the remarkable jointing phenomenon of this stone. The sundial is Irish in design and is one of only two found in Wales.[5]

These stones, like the churches, many of them are associated with, were once part of ancient *llannau* and would have sacred

5 The other sundial is in St Cadfan's Church, in Tywyn, a coastal town on Cardigan Bay in the South of Gwynedd (Edwards, 2013). Interestingly, it is possibly constructed of the same stone.

Figure 3.10 The 10th century sundial at St Beuno's Church is constructed from local Gyrn Ddu Porphyry.

significance. These religious sites have been incorporated into more recent churches, built from the 13th century onwards. Nevertheless, a small number of churches with clear pre-13th century structures and architecture still survive.

3.9.1 Early churches

Early churches predominantly dating from the 12th century and often occupying ancient *llannau* are found throughout the region. These are often associated with 6th–9th century inscribed stones described above, either standing in churchyards or incorporated into the fabric of the buildings which indicate much earlier uses of these localities. These churches are generally built of rubble masonry although many have been restored or rebuilt to some extent during the 19th century. This activity was often overseen by the ecclesiastical architect Henry Kennedy (1814–1896), who lived in Bangor for most of his life and worked on commission to the Diocese. Most of Kennedy's reconstructions were sympathetic to preserving original features, and he always reused the original stone where possible. The examples given here show a range of building materials from across the region as well as good examples of early Christian *llan* enclosures and inscribed stones.

Lligwy Chapel was built in the first half of the 12th century and is one of the oldest Christian stone buildings in the area under study (Figure 3.11). Now without its roof and part of the archaeological site that includes the Iron Age settlement of Din Lligwy and the Lligwy burial chamber, it is built from roughly coursed masonry which comprises very roughly hewn ashlar. These stones are a mixture of local Loggerheads Limestone and coarse-grained blocks of Basement Beds gritstone, which would also have been locally sourced. Some contain conspicuous, large, bright red fragments of jasper. These units outcrop well in nearby Lligwy Bay. A vault at the eastern end of the chapel probably dates to the 16th century and is roofed with slabs of Loggerheads Limestone. Inside the Chapel stands a roughly hewn stone block which would have been the base to a cross. This once stood inside the churchyard but has been moved inside for better protection. This base is also made from the Basement Beds, but unlike the facies used in the construction stone of the Chapel, it is a medium to coarse-grained, well-sorted sandstone.

Although dating mostly from the 12th century, and rebuilt in the 19th century by Kennedy, **St Cristiolus's Church** in central Anglesey was built on the site of an early Christian church and *llan*, and the church and its churchyard sit on this well-defined, sub-elliptical mound which is located 1.5 km east of the main village of Llangristiolus. Like Lligwy Chapel, it is built from the standard Anglesey construction materials in the form of local limestones (probably Loggerheads Limestone from Penmon or Moelfre) and Basement Beds gritstones and sandstones. Llangristiolus is also the burial place of the geologist Edward Greenly, who was the first to produce an accurate geological map of the island of Anglesey. The geology of his tomb will be discussed in the final chapter.

On the eastern end of the Llyn Peninsula, **St Aelhearn's Church** in Llanaelhearn was originally built c. 1200 from coursed rubble masonry. The village sits directly below the Iron Age Hillfort of Tre'r Ceiri, and this church too was built on a *llan*, the eastern edge of which, forming the boundary of the churchyard, is clear today. The

Figure 3.11 Lligwy Chapel is built of Loggerheads Limestone.

church is a low, cruciform structure which again was much restored by Henry Kennedy in 1848 (Haslam et al., 2009). An inscribed block of Lower Carboniferous Basement Beds sandstone with the date 1622 is set under the south window, indicating some building work in the 17th century. The church contains two early Christian inscribed stones including the 6th century Aliortus stone as well as a 6th–7th century stone pillar located outside the doorway (both described above). This church is of geological interest as it has incorporated into its fabric, over the last 900 or so years, a wide variety of igneous rocks from the Llyn Peninsula. The main stone used is a pink-weathering felsite porphyry. On fresher surfaces, texturally this stone is very distinctive with white feldspar phenocrysts set in an aphanitic, dark grey, almost glassy groundmass. This stone has been quarried from Mynydd Nefyn in the central Llyn Peninsula and also closer to Llanaelhearn from the Gyrn Ddu Pluton near Clynnog. In both locations it is well jointed with orthogonal joint sets and therefore breaks naturally into orthogonal slabs and blocks. Local felsite and felsite porphyry from the Tre'r Ceiri screes occur in the rubble masonry along with pyroclastic rocks from the Padarn Tuff Formation and metasomatised green shales from the Nant Ffrancon Group, which show a strong conchoidal fracture. This latter lithology is extremely localised and outcrops around Llithfaen to the south west of Llanaelhearn. Here the typically grey Nant Ffrancon Group silts have been modified by fluids associated with the Garnfor and Tre'r Ceiri Intrusions (Roberts, 1979). Some very large blocks and slabs of igneous are used in the construction of the church, particularly in the foundations, and an almost 2 m long quoin of porphyry appears on the west front. Original door arches are constructed from fans of porphyry slabs. Window dressings are yellow sandstones from the Anglesey Basement Bed gritstones (Davies, 2018b). Kennedy's rebuilding is clear on the eastern end where coursed ashlar masonry in pink and grey Trefor Granodiorite is used. This stone is quarried just 2.5 km to the north west of the village.

The old church of **Llandanwg** is located on the coast in the dunes, 3 km south of Harlech at the mouth of the Afon Artro. Situated on its *llan*, Old St Tanwg's had been abandoned and was becoming buried by dunes when it was excavated and became protected in 1884, when it was also given a new slate roof. The church was built in the 13th and 14th centuries; the Gothic East window belonging to this later phase. Both the church and the lychgate are built from uncoursed rubble masonry of glacial cobbles, derived from the hinterland of Snowdonia and therefore are largely Ordovician clastic sedimentary rocks and volcanic rocks (Davies, 2011). Window dressings are in Egryn Stone (Davies, 2011). The East Window has been partially blocked up and an early 15th century window inserted into it. This frame with three panes is constructed from coarse Basement Beds sandstones which Davies (2011) identifies as coming from Y Foel and Penmon on Anglesey. The doorway in the west front has a fan of slates forming the arch. These are from the local, Cambrian Llanbedr Slate Formation. A blocked up door in the south wall has a similar, fanned slate arch.

Old St Tanwg's church is also famous for its Early Christian Period stones, all now located inside the church. Three fifth to sixth century grave markers commemorating individuals named Gerontius, Ingenus Barbius and Equester are made of local Cambrian Hafotty Formation sandstones (Horák, 2013; Edwards, 2013). The Gerontius stone is split along bedding planes and shows the laminated character of this unit well. A mythology has developed that this stone came from Ireland, but the geological analysis confirms that it is indeed local. An incised cross, dating from the 7th to 9th century, is of coarser-grained Rhinog Grits, as is a cross-shaped incised stone of the 9th–11th centuries. All of these stones are derived from the local outcrops of Cambrian clastic sediments of the Harlech Dome.

3.9.1.1   Penmon Priory

The religious complex at Penmon, on the north east tip of Angle-sey, was established in the 6th century by Seiriol, who became one of the founding fathers of the Welsh Celtic Church and later be-came a saint. The main building, Penmon Priory is more correctly known as the Priory Church of St Seiriol (Figure 3.12). Evidence of an Iron Age Romano-British village at nearby Parc Dinmoor has been described above which may have been occupied up until the 5th or even 6th centuries CE. However, the only structure securely dating to Early Christian Period within the priory complex is St Seiriol's Well (Ffynnon Seiriol) and traces of a hermits cell, though arguably only foundations of this date survive. The brick and plas-ter well housing preserved today was built in the 18th century. The well itself is built into a quarry face of Cefn Mawr Limestone, the natural spring issuing from the stratigraphic contact of the Cefn Mawr Limestone and the underlying Loggerheads Limestone. There is little evidence of the quarrying of limestone in the early Medieval period, and it is perhaps surprising that limestone is not widely used in the construction of the Priory. As discussed above, one may speculate that part of this quarry site may indeed date

Figure 3.12 Penmon Priory. The Priory Church of St Seiriol is on the right. The building on the far left is the kitchen (warm-ing room) with refectory and dormitory behind. The main building material used here is Basement Beds gritstones and sandstones.

to the Roman period. The well is surrounded by slabs and seating constructed from slabs of Gwna Group, chlorite-muscovite schist which is quarried locally.

The Priory itself is an important and probably the finest example of Romanesque architecture in North Wales. Of the present Priory building, the cruciform nave, tower and transepts were built in two phases during the mid-12th century. The Chancel was constructed around a century later and a Prior's House, Refectory and Dormitory (now a private house) was added in the 13th century. In the early 15th century St Seiriol's priory was taken over by the Augustinian order and was further extended in the 16th century with the building of the kitchen, the so-called warming room, as an extension of the south wing. The church and priory complex were fully restored in 1855 by the architects Weightman and Hadfield (Haslam et al., 2009; Rcahmw, 1937). The south facing wall of the 19th century Chancel includes reused stone including fragments showing Romanesque moulded decoration (Figure 3.13). The complex of buildings are primarily constructed from Basement Beds gritty sandstones and conglomerates, quarried locally. The newer part of the church is Victorian, and the local Penmon marble is used for dressings and steps in the 19th century chancel.

In the nave of the Priory Church a *lapidarium* displays a number of carved stone crosses and other architectural elements which also date to the early Christian period. These include spectacular, 11th century arcading preserved in the South Transept with chevron moulded round arches. Figurative carving including carved heads and a Sheila-na-gig are probably from the same period as well as a 10th–11th century font with faces carved in a labyrinthine design and 10th–11th century Celtic crosses from Parc Dinmoor. All of these objects are carved from locally derived Basement Beds sandstones and gritstones (Horák, 2013; Edwards, 2013). Again it is an interesting choice given the fact that sharper and more-easily carved mouldings could have been made using the local limestones. The impressive Dovecote, standing to the east of the priory, was probably built much later, at around 1600. It too is constructed from predominantly Basement Bed gritstones and sandstones, whilst Penmon Marble (Loggerheads Limestone) is used to construct the corbelled stone roof.

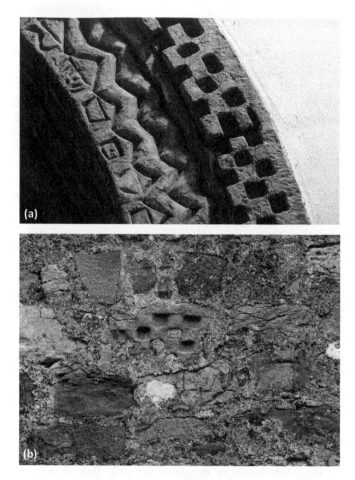

Figure 3.13 (a) Stone-carved, moulded Norman archways in the interior of The Priory Church of St Seiriol and (b) Fragments of the same decorative scheme used as rubble masonry on the west-facing wall of the church.

3.10   THE EARLY MEDIEVAL CASTLES

In 1066 William, the Duke of Normandy in western France, invaded England, defeating the English King Harold Godwinson at Hastings, and put Britain under Norman rule. By the 1070s, the Norman Earl of Chester, Hugh had his eye on the North Wales and particularly the fertile and profitable arable landscape of Anglesey. Over the next two decades he built a series of motte and bailey, earthwork mound castles along the North Welsh coast and thereby brought the word *castell* into the Welsh language (Butler, 2010).[6] These buildings included the precursors of what would in the later Medieval period become stone castles. Along with many other locales, a Norman motte was built on the future site of Caernarfon Castle and on the ancient Roman fort of Tomen-y-Mur. However, this incursion of Norman rule into North Wales was relatively short-lived. In 1,100, King of Gwynedd, Gruffudd ap Cynan evicted the Normans, ably assisted in this endeavour by the timely arrival of a band of Viking warriors and put himself on the throne of Gwynedd. Gruffudd's claim to the throne was legitimate, and he went on to form good diplomatic relationships between his court and the Norman English rulers. Gruffudd was succeeded by his son Owain in 1137, who was the first to describe himself as Prince of Wales and went on to establish a succession of monarchs up until Llywelyn ap Gruffudd, the last king of Wales (d. 1282).

It was during this period that the land was divided up into a system of administrative districts called *cantrefs*, each representing (approximately) one hundred towns or settlements which remain important in the administrative division of NW Wales today. The area covered in this book lies in the cantrefs of Rhosyr (on Anglesey), and Arfon, Arllechwedd and Dunoding. Each cantref was further subdivided into two commotes, and each commote contained a royal demesne, a *llys* or hall (Longley, 2010). It seems that the Princes of Wales used these royal residences peripatetically, travelling from one to another, rather than as permanent residences. These buildings were probably predominantly timber structures built on stone foundations as at the Rhosyr *llys* near Newborough on Anglesey (Johnstone, 1999) and unlike castles, they were undefended. During the Edwardian invasion and subsequent occupation, there is evidence that these buildings were dismantled and records of their timber being reused for the construction of castles, particularly Caernarfon is recorded in the building accounts (Piers, 1916).

---

6 The word *tomen*, meaning 'mound' was also already in use.

The building of earthwork castles began during the 12th century. Some of these occupied the sites of Iron Age hillforts and in some cases it has been difficult to disentangle the architecture of the Iron Age from that of the early Medieval period (Butler, 2010). The 13th century stone castles of Dolwyddelan, Dolbadarn, Criccieth and Deganwy were built by Llywelyn ap Iorweth (Llywelyn the Great, *Llywelyn Fawr*, b.1173 and the grandfather of Llywelyn ap Gruffudd), who reigned as King of Gwynedd and self-declared Prince of Wales from 1195 to 1240. These castles were built between 1220 and 1230 and were the first stone castles constructed in the region. It is clear from these monuments that Llywelyn the Great had more than a rudimentary understanding of castle building, these buildings are sophisticated in their design and far more than just forts. All are sited on rocky outcrops, and these underlying strata have also furnished these castles with their stone. Despite this, they are all relatively small structures and could never have housed a large garrison of armed men; they were little more than watchtowers. Nevertheless they would have had a commanding and authoritarian effect on the local region and its population. The rocky hills that form the foundations have allowed the castles to have not only a defensive position but also, and importantly, made them very difficult to undermine by tunnelling (though this method of attack on Deganwy Castle did prove to be its downfall). Dolwyddelan and Dolbadarn Castles guard mountain passes, whereas Criccieth and Deganwy overlook strategic maritime routes and ports.

3.10.1   Dolwyddelan Castle

Strictly speaking, Dolwyddelan Castle lies on the margins of the area covered in this book. It is situated in the Lledr Valley halfway between Bettws-y-Coed and Blaenau Ffestiniog. However, it is included herein as it was the first castle built by Llywelyn ap Iorweth and it was constructed in the village that is traditionally known as his birthplace. The overall plan of Dolwyddelan Castle is of a two storey, rectangular keep and a bailey enclosed by a curtain wall (Avent, 2010a). These sections were built in the early 13th century, probably around 1220. There are the remains of a second tower on the western edge of the bailey which was constructed in the late 13th century under Edward I. Maredudd ap Ieuan made repairs and further modifications were made to the keep in the late 15th century which included the addition of a new storey. The keep was further 'repaired' and remodelled in the mid-19th century (1848–1850) under the guidance, or perhaps misguidance of Peter Drummond-Burrell, 22nd Baron Willoughby de Eresby (1782–1865), who owned the Gwydir Estate in which Dolwyddelan sits. The walls were raised to a greater height and merlions were added to the battlements. However, the stone used was worked locally.

Geologically, the castle is located on the northern limb of the tight to isoclinal fold of the ENE-WSW trending Dolwyddelan syncline. This structure is periclinal and can be traced for 12 km. The northern limb is overturned, and all beds dip steeply to the north west (Howells & Smith, 1997). The core of the fold exposes the black Dolwyddelan Slates of the Upper Ordovician Nod Glas Formation and also the grey slates of the Cwm Eigiau Formation as well as the overlying Upper Rhyolitic Tuff Formation of the Snowdon Group volcanic rocks. The castle itself stands on a rocky knoll on the ridge defined by the hard, weather-resistant tuffs which form cliffs around the castle rock, allowing it to defend the valley below. This outcrop is also in part bounded by approximately N-S trending normal faults. The valley area below is formed over the subcrop and outcrop of the relatively much softer slates.

The tuffs and the slates are the two main stones used for the construction of the keep and its walls throughout the history of building and rebuilding at Dolwyddelan Castle (Neaverson, 1947). The grey-green Upper Rhyolitic Formation tuffs are the main building stone applied in coursed rubble masonry. The tuffs break into roughly rectangular slabs along bedding, joints and cleavage planes. Thomas Pennant (1726–1798) toured Wales in the 18th century and wrote of Dolwyddelan Castle 'the materials of this fortress are the shattery stone of the country, well-squared, the masonry good and the mortar hard' (quoted in Barnwell, 1883), a fitting description of these somewhat intractable building materials. The black slates are also used as slabs and fillers within the rubble masonry, but are predominantly and effectively used as dressings, particularly around the arches of doorways and windows, where they are employed with the prominent cleavage orientated orthogonal to the line of the arch giving a fan-like impression, with a wedged-shaped block of tuff in the middle, forming a keystone. 'Dolwyddelan Black Slates' were also quarried locally as roofing material. The earliest known slate quarry in the Dolwyddelan district is Chwarel Ddu (the Black Quarry) which lies just below the castle rock. This was opened in

1800, but it may well also have been the source of the small amounts of stone used in the Medieval period, any evidence of earlier use having since been quarried out. A third stone has been identified as being used in the castle, and this is not local to the site. The West Tower has elements, including the threshold stone and some window dressings of a mottled purple-red and white sandstone Neaverson (1947), which is probably Gloddaeth Purple Sandstone from the Conwy area and used in Conwy Castle; this unusual lithology will be discussed in more detail in the following chapter.

3.10.2   Dolbadarn Castle

Dolbadarn Castle sits on a rocky hillock on the isthmus which separates Llyn Padarn and Llyn Peris in what is now the village of Llanberis (Figure 3.14). It is immediately below the huge 19th century quarries of Dinorwic from which the Global Heritage Stone Resource of the Welsh Cambrian Slate is worked (Hughes et al., 2016). In the 18th and 19th century, the steep-sided, U-shaped, glacial valley of Nant Peris became a favourite landscape of painters of the sublime, the perfect view of a castle with the lake below it and the massif of Mount Snowdon behind it, dropping down

Figure 3.14 Dolbadarn Castle is built from slabs of slate from the Llanberis Formation.

to the narrow cleft of Llanberis Pass beyond. Dolbadarn castle was constructed to defend and dominate the Pass which then and now enables access across the highest region of Snowdonia and down to the coastal plain of Arfon. It was built by Llywelyn ap Iorweth in the early 13th century, probably c. 1220–1230 and it is unique amongst his Welsh castles in that its keep is a round tower (Avent, 2010a). The keep at Dolbadarn had three storeys and the entrance was via a mural stair which circles around the exterior of the keep. The keep stands in the curtain walls which enclose a small bailey, the size and shape of this circuit of walls are entirely defined by the shape of the hill. Rectangular towers, now reduced to foundations, are situated at the south and west, and a great hall and other buildings are located on the north and east sides of the bailey. All structures, with the exception of the keep, have been reduced to foundation level.

Dolbadarn Castle is built on an outcrop of the Llanberis Slate Formation and the keep and its walls are almost entirely built of this stone (Neaverson, 1947). Indeed the slates outcrop prominently within the bailey; levelling of this area was clearly not a priority in these fortresses. All facies of these multicoloured slates are used in the building. Green, blue and purple slates, the latter with prominent reduction spots can all be identified within the fabric. Slabs of slate, extracted along cleavage planes, lend their aspect ratio to well-coursed rubble masonry. Some window arches use fans of slates with cleavage orientated orthogonal to the line of the arch, as described at Dolwyddelan above; however, others use large slabs of slate as both lintels and sills. Slate is also used as paving within the castle compound. Slate slabs, secured by iron clamps, are used to cap the stones of the wall of the stairway, though these are probably modern additions. The only exceptions to the use of slate are a few rounded boulders from the local glacial deposits, no doubt picked up on site or on the lake shore, which are also used as rubble fill in the walls.

3.10.3 Deganwy Castle

Very little remains of Deganwy Castle at the present day. On casually driving past the low, rocky summit on which it was built, one has to know that the masonry is there to notice it. Deganwy was built across two prominent outcrops of the rhyolitic tuffs of the

Ordovician Capel Curig Volcanic Formation which overlook the estuary of the River Conwy from its north bank in the commote of Creuddyn. The castle was also built of the tuffs. The earliest castle on this site was one of the Norman mottes built by Hugh of Chester in the 11th century. Llewelyn the Great built the first stone castle on this site in the 1230s. However, this fortification was captured and expanded by Henry III between 1247 and 1249, work that included the addition of a new tower. In 1263, the castle was besieged, recaptured and subsequently slighted by Llewelyn ap Gruffudd (Nevell, 2020). The castle was excavated in the1960s by Leslie Alcock (Alcock, 1968), who found considerable evidence of destruction, which appeared to be related to slighting rather than to damage incurred by the castle being besieged. Intentional destruction included tunnelling under Henry III's tower in order to cause its collapse. Similarly some 450 m of the curtain wall was undermined and destroyed and a large quantity of rubble masonry was found in the ditch on the south side of the bailey. The main building stone was the buff yellow to slightly greenish tinged and strongly iron stained acid pyroclastic rocks of the Capel Curig Volcanic Formation. It is possible that some of the stone from Deganwy Castle was reused for building the town walls of Conwy in 1283, but this cannot be proven. Certainly a lot of stone was not removed from the site and remains *in situ*.

The most famous find from the rubble of Deganwy Castle is a finely carved head of a king, thought to be a likeness of Llewelyn ap Iorweth[7] (Avent et al., 2011) and carved from a micaceous, well-sorted, medium to coarse-grained, quartz sandstone. This is assumed to be a Carboniferous sandstone sourced from north east Wales. There are a number of very similar sandstones in this region, and similar stone heads of similar ages such as one found at Valle Crucis Abbey near Llangollen[8] have been identified as stone from the Cefn-y-fedw Formation of the Millstone Grit Group. The presence of mica in this head suggests that it is more likely to be Cefn Sandstone (not to be confused with Cefn-y-fedw sandstone) which belongs to the Pennine Middle Coal Measures Group. Lott (2009) and Waters et al. (2007) have examined the petrology of these lithologies and conclusively demonstrated that they contain muscovite.

---

7 National Museum of Wales Catalogue Number 77.11H/10.
8 National Museum of Wales Catalogue Number 77.36H/10.

3.10.4   Criccieth Castle

Criccieth is located midway between Caernarfon and Harlech in the commote of Eifionydd, at the south eastern end of the Llyn Peninsula (Figure 3.15). It is probably the most impressive of Llywelyn the Great's castles and was probably influenced by the architecture of Beeston Castle (built c. 1220) which stands on a prominent hillock on the Cheshire Plain (Avent et al., 2011). Criccieth Castle owes its lofty position to a stock of pink-coloured, finely-porphyritic, igneous rock of rhyolitic composition, one of the Ordovician Nefyn Cluster of intrusions. It is probably best described using the field term 'felsite'. A precise age for this stock and its associated outcrop nearby at Dinas, on the opposite side of the Castle Ditch, is not available. Exposures on the south east face of the Castle Rock and the south west facing, quarried face of Dinas show the felsite to be well-jointed and therefore probably relatively easily quarried and cut into usable blocks. A prominent depression on the north facing side of the Castle Rock, immediately below the modern entrance up to the castle was probably one of the main quarries. The first stone Castle on the site was built by Llywelyn the Great between 1230 and his death in 1240. This structure is now the inner ward of the castle and retains its impressive gate house with twin, D-shaped towers on the eastern side, now termed

Figure 3.15 Criccieth Castle is built in rubble masonry from the felsitic intrusive plutonic rock on which it stands.

the Inner Gatehouse. The castle was further fortified with a circuit of walls and towers (the South-West and North Towers) which enclosed the outer ward of the castle which was constructed by Llywelyn the Great's grandson Llywelyn ap Gruffudd during the 1260s to early 1270s. After the death of Llywelyn in 1282, Edward I occupied the castle and further fortified the North Tower and built the South-East Tower. In 1404, the castle was taken during Owain Glyndwr's uprising and set on fire. Burnt destruction debris was found during early excavations at the Castle under the direction of St John O'Neil (O'Neil, 1944).

The main building stones used in the Castle are the pink felsites from the Castle Rock and Dinas. The wall cores are of rubble masonry bonded with lime cement. Indeed in the Exchequer accounts during the rebuilding, the main expenditure was for the carrying of lime mortar and sand, supporting the evidence that all stone was locally sourced. The walls are faced with roughly coursed ashlar masonry. Dressings around windows and arrow loops are somewhat rough and ready but include of slabs of dark grey.

Ordovician sandstones of the Dol-cyn-Afon Formation (Mawddach Group) sourced from the vicinity of the present day town of Porthmadog. The North Tower (also called the Engine Tower) is almost entirely built of slabs of this sandstone and is replete with slots for a range of latrines (Avent et al., 2011). Dressings around doorways in the Inner Gatehouse and in the South East Tower are constructed from the Basement Beds Lower Carboniferous sandstones of Anglesey along with some channelled stones used for water supply and plumbing. In all cases the medium-grained, buff-coloured freestones of the Malltraeth facies have been used (Shipton, 2020; Neaverson, 1947). A few small pieces of slate are also not infrequently encountered in the castle walls. Evidence of the burning of the castle, in the form of the reddening of the pink felsite, is apparent in several places, but most obvious in the South East Tower.

In addition to the extant architecture, St John O'Neil's excavations recovered the carved head of a woman in 'pinkish sandstone', some elaborate stone chimneys and a number of rotary querns, all of Basement Beds gritstone. F. J. North (Keeper of Geology, NMW) examined the stones and identified all sandstones to have been derived from the Basement Beds of Anglesey and writes 'I imagine that there would be no need to look further afield for their place of origin' (O'Neil, 1944). The samples of stone O'Neil collected from Criccieth Castle are still in the collections of the National Museum in Cardiff.

3.11   THE LAST KING OF GWYNEDD

Following Llywelyn the Great's death in 1240, he was briefly succeeded by his son Dafydd ap Llywelyn and eventually by his grandson Llywelyn ap Gruffud, also known as Llywelyn the Last (*Llywelyn Ein Llyw Olaf*) in 1246. The power of the Welsh princes was in decline and from 1247, incursions into their territories by the English King Henry III began. These border skirmishes and taking of land marked the beginning of the end for Welsh rulership. Henry's son and heir, Edward I, a man who's attitude to war was more focused than that of his father, invaded North Wales first in 1277 and then decisively in 1282. Llywelyn ap Gruffudd was killed in the wars, and his successor and brother, Dafydd was captured and then executed the following year. Edward I secured and sealed his dominance over the North Welsh landscape and its peoples by constructing a series of new stone castles with adjacent new towns which surpassed all previous standards, in scale as well as military and defensive architecture.

Chapter 4

The castles and town walls of Edward I

4.1 INTRODUCTION

The two English Kings whose reigns spanned much of the 13th century were both great builders though otherwise, seemingly of very different dispositions. Henry III (reigned 1216–1272) devoted much of the later part of his life to rebuilding Westminster Abbey and designating it as a shrine for St Edward the Confessor. He named his eldest son Edward after the Saint. Henry's reign was one of relative peace until he needed to raise money to pay for the lavish buildings works at the Abbey and to replenish the exchequer; these events precipitated the Second Baron's War. His son, who then went to war on his father's behalf, seemed always to have had a fight on his hands. His building projects, including the four great castes of North Wales, now designated a World Heritage Site, were constructed to subdue the people of his conquests in Wales.

All of Edward's castles are built on defensive and strategic sites; Caernarfon and Beaumaris guard the entrances to the Menai Straits, Harlech's lofty position commands views of a large proportion of Cardigan Bay and Conwy sits at the mouth of the navigable River Conwy. However, defending the Welsh coast from raiders or enemy fleets was never really the intention of these structures at the time of their building; their role was arguably more offensive than defensive. They were much more about symbols of power, superiority and dominance rather than means of protecting the Welsh coasts and their hinterland. Nevertheless their coastal positions meant that they were easy to supply with both building materials and consumables and also made it possible for the constables to collect taxes from goods coming into port on behalf of the King.

DOI: 10.1201/9781003002444-4

Edward was born in 1239, the eldest son of Henry III and Eleanor of Provence. Edward married Eleanor of Castile in 1254, and with that marriage contract he acquired the Earldom of Chester which included land in North Wales between the Rivers Dee and Conwy comprising four *cantrefs*. His father, Henry III was king of England, but much of Wales was a principality under the rule of the King of Gwynedd Llewelyn ap Gruffudd. Henry's reign had been largely peaceful, but towards the end of his life, he began to bring in reforms of tax and land rights and subsequently both King and heir fell foul of the barons. One man's reformer is another man's tyrant. Mismanagement of the four North Welsh *cantrefs* by Edward's officials in Chester caused local rebellion and an uprising was led by Llewelyn. A number of skirmishes and battles were fought during 1254 and 1263, with victory largely on the side of the Welsh Prince. But by 1263 Edward had a battle on another front. A long-standing dispute between Edward's father, King Henry III and Simon de Montfort, sixth Earl of Leicester, finally erupted into civil war. De Montfort quickly allied himself to Llewelyn but was then defeated in battle. Following the war, Llewelyn successfully negotiated the Treaty of Montgomery with Henry III which firmly established him as Prince of Wales and ceded all Welsh lands to him, including the *cantrefs* that were part of the Earldom of Chester. The loss of these lands would go on to cause tension between Edward and Llewelyn and eventually led to outright war.

In 1270, Edward went on Crusade and was conveniently away from his trials in Britain for 4 years. Henry III died in 1272, and Edward succeeded the throne *in absentia* but was not crowned until he returned from the Near East in 1274. The bitterness with Llewelyn had not dissipated and Llewelyn refused to attend the coronation and also, to further taunt Edward, he had also planned to marry Simon de Montfort's daughter. Provoked by this behaviour, Edward took a substantial army to Wales with the intention of reclaiming his four cantrefs and subduing Llewelyn. Indeed, Llewelyn was defeated and grudgingly agreed terms with the English King. However, quarrels between English and Welsh lords continued, and in 1282, Llewelyn and his brother Dafydd raised an army and attacked Hawarden Castle in Flintshire which was then in English hands. Edward responded swiftly and vengefully, raising an army to invade Wales and bring with him a team of engineers to construct both castles and a pontoon bridge across the Menai Straits (the 'ponte meney') to enable his army to access Anglesey, the agricultural heartland of North West Wales. War commenced. Llewelyn was killed in December 1282, and his brother Dafydd was captured and executed in 1283. Edward transferred all Welsh lands to 'his dominion' (Statutes of the Realm quoted in Prestwich, 2010). The war of 1277 initiated the building of the castles of Fflint and Rhuddlan whilst Caernarfon, Harlech and Conwy castles began construction in 1282. By this time, Edward I was engaged in yet another war, this time abroad with Gascony. The Welsh took advantage of his absence, and rebellions occurred in 1287 and again more successfully in 1294. This latest affront to his power provoked Edward into taking another army to North West Wales, and it was during this campaign that he began construction of Beaumaris Castle on the isle of Anglesey.

War with Scotland in the early 14th century then took Edward's battles to another front, and by 1307, he was dead from dysentery, contracted on his campaigns against Robert the Bruce. He was succeeded by his son Edward of Caernarfon (King Edward II). The building of the North Welsh castles continued for the next decade or so, but with Edward I gone, some semblance of peace was regained in the region. The castles continued to be garrisoned until the early 15th century. When Henry Tudor, a man with Welsh ancestry, came to the throne as Henry VII in 1485, relations between England and Wales improved significantly, and the need for either defensive or indeed offensive castles became less important, and for the next 100 years or so, the castles fell into decay. The Castles last saw military action and assumed a new defensive role during the English Civil War of 1642–1651. The castles, like many others in the British Isles, were once more garrisoned and repairs, especially to rotten timberwork, were necessary. At the end of the war, the Welsh castles were slighted,[1] though outwardly at least, it appears they survived this destruction better than many in England. Following the war, the castles stood large empty, probably occupied by squatters, but with little in the way of repairs or maintenance. During the late 18th century, the Romantic movement brought these crumbling but picturesque ruins by the sea to a new audience. The English

---

1 Slighting is the deliberate destruction of castles, often undertaken for political reasons, with the intent of reducing the value of a building and its associated region and consequently limiting or destroying the power of local aristocracy (see Nevell, 2020).

artist J. M. W. Turner (1775–1851) visited North Wales several times in the 1790s and made paintings of all four of the Edwardian Castles. Turner and other artists increased the popularity of the ruins and during the 19th century, tourism put the castles back on the map. This was particularly the case for Caernarfon Castle, located in the local administrative centre and town which was growing into a busy slate port.

In the early 19th century, the Edwardian Castles were being used as munitions dumps and as warehouses from goods coming into the ports. Repairs were made only when they posed a risk to the public and the buildings and thoroughfares beneath their walls. Harlech, which was never a large town, became isolated and (until the arrival of the railways) cut off from the north west coast of Arfon and the Llyn Peninsula. Caernarfon, Conwy, Harlech and Beaumaris Castles were cleared of debris and vegetation during the 19th century, but the largest and arguably most impressive castle, Caernarfon found itself much more the limelight becoming the sight chosen for the investiture of the Prince of Wales in 1911. All the castles are now in the care of Cadw, the organisation which acts for the preservation of Welsh historic monuments on behalf of the Secretary of State for Wales. The four Edwardian Castles and their walled towns were designated World Heritage Sites in 1986.

Today, and over the past century, the Castles of Edward I still provoke mixed feelings amongst Welsh population. Despite being some of the best examples of Medieval military architecture in Europe, they are seen by some Welsh people as offensive buildings; symbols of defeat, invasion and subsequent oppression of the local population. This has led to protests at the investitures of the Princes of Wales in both 1911 and 1969 and also a backlash to Cadw's proposed 'celebration' of the 700th year anniversary of the main phase of castle construction in 1983. Similarly a petition was signed in 2017 to prevent the installation of a sculpture in Fflint, the 'Iron Ring' which symbolised the defensive unity of Edward's castle-building campaign (Jones, 2010; Sands, 2017).

4.2 BUILDING CASTLES AND CASTLE BUILDERS

Although there had been numerous castles built in North Wales and elsewhere in the British Isles, castles the on the scale of Caernarfon, Conwy, Harlech and Beaumaris had not been seen in Britain before

Edward I's reign. Much of the responsibility for constructing these astonishing examples of military architecture must lie with the experience, knowledge and foresight of the King himself. Edward would have developed some considerable knowledge of the practical construction of fortifications after his experiences of warfare and journeys in the Holy Land and the French region of Savoy.

From his experience of wars in the Welsh Marches, battles with Simon de Montfort and the barons and his crusade journey across the Europe and North Africa, Edward had seen many castles and was well aware of their strategic and defensive advantages. During his Crusade he had certainly travelled as far as Acre (in modern-day Israel) where he was involved in building a new tower to the walls of the citadel there. The Crusader Castles of the Levant are legendary, although Edward probably never saw the great crusader castle of Krac des Chevaliers,[2] which had fallen in 1271, before he arrived in the region. Nevertheless his journey to the Holy Land had taught Edward that castles could be supreme, fortified structures with the power to dominate a region. As Fedden and Thomson (1957) remark, 'castles in themselves cannot prevent an invading army marching where it pleases, but if conquest is to be permanent, the castles must eventually be taken'. It would also have been in the Holy Land that Edward learned of the offensive nature of castles, whereby local populations could be subdued. Such castles blockaded ports and thus controlled food supply and revenues from other goods, performing multiple and complex roles as a military garrison, customs house, prison and royal palace.

From surviving letters and building accounts, it is clear that Edward played a major part in the design of the castles and spoke directly with his top men on site, namely Master James of St George and Richard the Engineer (Richard Lenginour, c. 1240–1315). These two important names stand out in the construction accounts of Edward's castles. What is known about these two men has been discussed by Coldstream (2003, 2010) and Turner (2010).

Edward I had connections in Savoy, via his mother, whose uncle was Peter II, Count of Savoy (and the builder of London's Savoy Palace). On his return from the Holy Land, he had spent time in

---

2 Krac (Crac) des Chevaliers was built by the Knights Hospitallers in 1142 and remained in action until 1271. It was created a UNESCO World Heritage Site (No. 1229) in 2006.

Savoy and stayed at the Count's castle, St Georges d'Espéranche. It was surely here that he encountered the local engineer and architect Master James, later to be known as James of St George, possibly named after this castle. On crossing the Alps from Italy into France, Edward would have seen and may even have visited the Savoyard castles that protected the mountain passes, perched on precipitous slopes, some with round towers and some with octagonal towers. Similarities between the castles of Savoy and the Welsh castles have long been acknowledged (see Taylor, 1950); nevertheless those of Wales take the concept of castle architecture far beyond the fortifications of the Savoy, including the castles at Yverdon, St Georges d'Espéranche and Chillon on Lake Geneva. Important 'Savoyard' architectural and engineering features which particularly link the two regions are as follows: put-log holes for supporting scaffold which spiral around the towers; semi-circular arches in doorways; windows with tracery; castellations (merlions) with decorative pinnacles and two distinct types of latrine shafts, a semi-circular, corbelled variety and a shaft which projects a short distance from the walls (Figure 4.1). Whether these features prove

Figure 4.1 The south wall of Harlech Castle showing various aspects of Savoyard masonry. Note the spiralling putlog holes around the tower, and the two types of latrine shafts. A semi-circular, corbelled shaft can be seen in the lower part of the walls and a narrow, rectangular shaft can be seen projecting from the wall above this. The bedrock of the Castle Rock, the Rhinog Formation.

a direct link between the Savoy and North Wales has recently been questioned by Coldstream (2003, 2010), who points out that some of these individual features are not unique to these local architectural traditions.

Although he had worked as a mason and overseer in France, it is now known that Master James had not been the architect of any of the Savoyard Castles, but his abilities began to shine when in the employ of Edward. He first appears in the records in 1278 when he was first sent to work in Wales at the castles of Fflint and Rhuddlan (where work had started the previous year; there is no evidence to suggest that Master James had played a part in their design), and in the 1280s when Caernarfon, Harlech and Conwy were being built, he was named as Master of the King's Works. All evidence suggests that James learned much of his craft as builder-designer and overseer on the job in Britain (Coldstream, 2003). The work of Taylor (1950) and subsequent publications cited in Coldstream (2010) showed that Master James brought with him a number of masons and builders, whose names occur in construction records and accounts in both Savoy and Wales; people such as Gilet of St George, John Francis and Beynardus. Many of these men, involved in castle building at home, did not come over to Britain until the 1280s, perhaps bringing with them new ideas from castles they had built which James himself could not possibly have seen (Coldstream, 2010). John Francis had worked at the Chateau de Saillon in what is now the Swiss Valais, a structure built on a rocky outcrop, with towers very similar in design to those at Conwy. Such reassessment of the timings of arrivals of the various builders and engineers dispels the belief that Master James had arrived in Wales replete with the architectural plans for the four new castles, built to Edward's specifications. It is more likely that the architectural development of the castles evolved through time with knowledge and expertise between British and French builder-architects, brought together into a master plan overseen by James of St George. Also of huge importance and as Coldstream (2010) neatly points out, Master James's great skill was the organisation of construction of this immense project in a wild and rugged landscape.

Master James was subsequently given more far-ranging jobs. The main phases of building of the four castles came to a close in around 1290; he was appointed constable of Harlech Castle, a role which he held until 1293, and he was later to assist Edward I on his Scottish campaigns (Ashbee, 2017a).

Another important name is Richard the Engineer, who was an Englishman from Chester and in 1265 is recorded as superintendent of works at Chester Castle for the (then) Prince Edward. He was probably a carpenter by trade but was elevated to supervisory and organisational roles on a number of building and engineering projects. He was clearly trusted by Edward and went on campaign with him into Wales in 1277. Richard began work as an overseer at Fflint and Rhuddlan Castles where part of his role was a responsibility to procure stone from quarries at Shotwick for the construction of Fflint Castle (red sandstones of the Permo-Triassic Kinnerton Sandstone and Chester Pebble Beds Formations). He later goes on to work at Caernarfon and Conwy with Master James and is also the man responsible for the construction of the pontoon bridges across the Menai Straits, the first at 'ponte meney' and the second from Bangor towards Llanfaes for the construction and garrisoning of Beaumaris Castle. He was certainly responsible for supervising digging of the stone ditches at Conwy (Turner, 2010), and one may therefore speculate that his experiences may well have given him the responsibility of procuring stone for the construction of these castles. What is clear is that Edward knew Richard personally, and in the accounts, his salary is second only to Master James's, suggesting that he had authority and even autonomy over certain works (Coldstream, 2003).

Sir John de Brevillard is named as supervisor of the works in the first phase of building Caernarfon and Conwy Castles in 1280s. He then went on to be Constable at Harlech Castle in 1285 but died 2 years later. Walter of Hereford was in charge of the second phase of building at Caernarfon Castle, which was initiated in 1295 (Coldstream, 2003). At Harlech Castle, we see the name of a master mason, William of Drogheda (in Ireland), who steps up when Master James's work under the King's orders temporarily takes him to Gascony in 1287. Also during 1277–1279, masons Master William and Hugh of Wem are recording as working on the eastern towers of Harlech Castle (Ashbee, 2017a).

Although the focus of this book is on the stone building, it must not be forgotten that wood was also an important construction material used in all the castles, for constructing roofs, internal stairs and free-standing buildings within the castle wards. Richard the Engineer has started life as a carpenter, and a Master Henry of Oxford is named as working at Conwy (Ashbee, 2006). The role of these men would also have included the construction and maintenance of scaffold, essential but temporary constructions in the building of any substantial building. The only remaining evidence for scaffolding is the put log holes which are seen in the castle walls. These features would have supported cantilevered wooden beams on which scaffold platforms and ramps could be constructed.

Also indispensable men on the construction projects were the Clerks of Works who had responsibility for the payment of bills and wages and would have worked closely with Master James and other overseers. Hugh of Leominster is named as Clerk at Caernarfon. Later at Beaumaris, the Clerk of Works is a Walter of Winchester, who was much troubled with managing the construction of a castle with rapidly diminishing funds (Ashbee, 2017b; Coldstream, 2003). These men and others like them were responsible for the payment of a huge workforce of masons, quarrymen, lime burners, plasterers, carpenters, scaffolders, blacksmiths, boatmen and general labourers, who ranged from women who collected moss on the mountains which was used to cushion the roofing slates to dock workers who unloaded building material. At Beaumaris in 1296, some 1,600 people are recorded working onsite. Women may have played a greater part in the skilled workforce than has been generally assumed; certainly women were named as operating lime-burning business at the time and also worked as smiths and carpenters. One of these was a Cecilia of Kent, who along with her female assistant was making parts for artillery machines. At Caernarfon Castle, women are listed as working with lime burners and plasterers as mortar carriers; Julia and Emmota, both carpenters' daughters and 'Ade's wife, Juliana' along with Eleanor of England are named in the accounts and paid equally to men working in the same roles (Piers, 1916; Lloyd Jones, 2016).

Mason's marks of fifteen different types were found on stone blocks in Beaumaris Castle. Most of these were discovered following a flood in 2015, following which the moat was drained, revealing the lowermost courses of walling (Lloyd-Jones, 2016). Reproductions of the marks have been recreated on sawn blocks of recently quarried Penmon Marble, and these have been placed around the castle to highlight the location of and make more visible mason's marks as part of the current public archaeology programme.

It is perhaps incredible to realise that Caernarfon, Conwy and Harlech Castles were all being built simultaneously, with the construction of all three starting in 1283, and all largely under the administration of the same group of people. These buildings were constructed incredibly rapidly; reaching operational status often with 2–3 years from the start of construction. Presumably building and labour teams at each location were unique, but this fact suggests that master engineers, masons and carpenters were mobile, travelling constantly between these three major construction sites. Building at Beaumaris began over a decade later in 1295. Despite a unified approach to their construction, all four castles have their own unique architectural character and building materials, the majority of which were sourced very local to the construction site.

The building materials of the Welsh castles have been of interest to geologists since the early 20th century. The geologist Edward Greenly was the first to make a survey of the building stones of Caernarfon and Beaumaris Castles (Greenly, 1932). A more thorough study, taking in the building materials and water supplies of all the North Welsh castles, including those built by the Welsh Princes prior to the invasion and conquest of Edward I, was subsequently undertaken by Neaverson (1947). More recently, Nichol (2005) has reviewed materials used at Caernarfon, and Lott (2010) has provided a more in-depth study of the main materials used at Caernarfon, Harlech, Conwy and Beaumaris. Their work has provided the foundations of the observations made herein and the building materials for each castle and where relevant, the adjoining town walls are described below. Caernarfon Castle, however, is unique in its size, architectural style and use of materials and will be discussed first.

4.3   CAERNARFON CASTLE AND TOWN WALLS

4.3.1   Caernarfon Castle

Caernarfon is the largest, most imposing and most architecturally distinct of Edward I's four castles built during his military campaigns in Wales during the last 20 years of the 13th century (Figure 4.2). It consists of curtain walls connecting ten octagonal towers, which feature stripes of coloured stone which provide contrast with the main building stone. Similarities in appearance of the castle walls to the Late Roman Theodosian Walls of Istanbul (Byzantium) constructed in the 5th century AD had not gone unnoticed by the historian A. J. Taylor, who made much of this perceived connection, linking the construction of Caernarfon Castle to Edward's experiences and observations whilst journeying on Crusade (Taylor, 1993). The walls of Byzantium feature octagonal

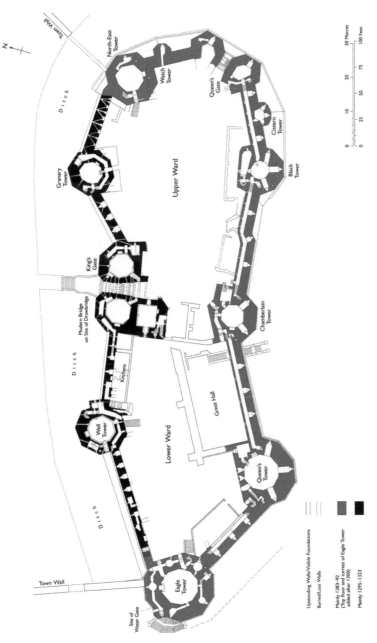

Figure 4.2 A ground plan of Caernarfon Castle. (Crown copyright Cadw (2020).)

as well as square-section towers separated by curtain walls, along which continuous stripes of contrasting brickwork ornament the walls (Figure 4.3). The stripes on the Theodosian Walls are numerous, at least ten stripes. However, it must be acknowledged that this decorative style in Roman architecture was not unique to Byzantium and many architecturally similar constructions exist, perhaps even very close to the site of the Castle at Segontium (Swallow, 2019; Wheatley, 2010). The influence of Roman architecture, as interpreted by Swallow (2019), may have been a desire by Edward and his Queen Eleanor of Castile to link their new palace-castle with the romantic folk tale 'The Dream of Macsen Wledig', a story which now forms part of the Mabinogion.

The origins of the story of Macsen's dream are complex but centre upon a mythical Roman Emperor Magnus Maximus, who dreams of a castle on a river mouth by the sea wherein dwelt a beautiful maiden Elen. Maximus's envoys travelled the world until they found Caernarfon and realised that it was the castle in the dream and the home of Elen. Macsen travelled to the castle by the sea and married Elen, and they both ruled the land of Arfon together until Maximus returned to Rome to defend his absentee imperial claim. This symbolism with ancient Roman royalty and romance of the past and particularly with the heroine being Eleanor's near namesake may have been a story that the royal couple literally built upon. There was a real Magnus Maximus, he was a Roman General who commanded the British garrisons who subsequently proclaimed him emperor in 383 AD. Swallow (2019) argues that the myth must have appealed to Edward and Eleanor, who may have seen themselves as the new Macsen and Elen, a foreign King ruling benignly over Arfon with his Queen at his side and that this may account for the prominence of the Queen's Gate in the Castle's architectural design.

Geographically Caernarfon Castle is built upon what would have been a rocky promontory on the Menai Straits located between the Rivers Seiont and Cadnant, a site once occupied by a (presumably) timber-built, Norman motte and bailey castle, and maybe also the site of a *llys* of the welsh Kings (Swallow, 2019).[3]

---

3 The Roman 'castle' of Macsen, is of course the Fort at Segontium, above the main town, and despite the dream, not on the riverbank. However, it is not impossible that there were wharves in this location during the Roman period.

Figure 4.3 (a) View of Caernarfon Castle from across the River Seiont. The fine masonry of this river-facing façade of the Castle is executed in limestone ashlar with four decorative stripes in yellow sandstone. (b) A view of a section of the Theodosian walls of Byzantium (modern Istanbul, Turkey). These walls constructed in the Roman Period (5th century CE) show a very similar masonry style to that seen in Caernarfon, including the use of octagonal towers. It is possible that Roman military architecture influenced Edward's design for Caernarfon Castle.

The Norman motte was located in what is now the upper ward of the Edwardian Castle, and the bailey would have extended eastwards into the area that is now the town square, Y Maes (Cadw, 2010). This area between the two rivers was occupied by the town which was established and designed concurrently with the Castle. Clearing the land, digging the castle ditch and building began in the Summer of 1283, with the early establishment of wharves to receive consignments of timber and stone brought in by sea. The King was present to oversee works during July and August, and the Royal family returned to Caernarfon for Easter 1284. Tradition has it that it was here that Edward and Eleanor's son Edward of Caernarfon was born on 25th April 1284, though there is little clear historical evidence to support this, outside the fact that the child was subsequently known as Edward of Caernarfon.

The main works on the first phase of the castle took place between 1285 and 1291. The town walls and the south curtain wall, facing the River Seiont, were the first to be completed. These walls would have been an impressive frontage to the new castle. They are faced in limestone and have a series of contrasting, horizontal stripes in a yellow to red sandstone. This feature is clearly intended to be decorative and therefore indicates that the castle walls were never intended to be rendered with plaster.

In 1285, Exchequer accounts record that workmen were diverted to dig out the King's Pond, a large fishpond, probably intended for storing trout. This was located just north of the main castle, and its former existence is recorded by the name of the main thoroughfare in Caernarfon town centre, Pool Street (Stryd-y-Pwll).[4] Little was spent on the Castle that year; 'boards, shingles, nails and glass windows' were the only expenditure at £6 3s. 4d. (Piers, 1916). It appears that these materials were being used to construct wooden apartments within the castle ward. Otherwise work appears mainly to have been concentrated on the town walls and the pond. These works appear to have been completed by 1286.

Building works continued until 1292 when they were bought to a halt by Edward I's war with Gascony. Edward's distraction

---

4 The pool was still in existence until the mid-18th century; it appears of John Speed's Map of Caernarfon published in 1611 and on J. Wood's map of the town of 1834. It was destroyed during the engineering works to create the railway in 1869 (Cadw, 2010).

overseas was seen as an opportunity by the local Welsh nobility, and in 1294, a rebellion led by Madog ap Llewelyn, a relative of Llewelyn the Great, burned the town, destroying parts of the newly built town walls and put a torch to all the wooden structures in the town and in the castle wards. This uprising was quickly suppressed, and Madog was sent to the Tower of London where he was imprisoned for the remainder of his life. Building at Caernarfon Castle resumed in 1295, repairing and reconstructing the damaged town walls and this work continued until the end of the century. By 1300, Edward was at war with the Scots and once again building work ceased. The final phase of building at Caernarfon Castle began in 1304 and went on for a further decade, by now under the direction of Edward of Caernarfon, who had succeeded his father as King Edward II. The town-facing curtain wall of the castle was completed, and the upper floors and three turrets were added to the Eagle Tower. This phase of building made certain that Caernarfon Castle would not be taken again. The new north wall contained a kitchen block, but more importantly a formidable and highly sophisticated military battery, which would have allowed a team of archers to make covering fire across the whole front of the castle whilst remaining protected. The final addition to the castle, in the early 1320s, was the completion of the turrets on the King's Gate and the installation of the statue of King Edward I on the front of the gatehouse. This was fixed to the wall using iron cramps and also required 'twelve iron spikes to protect the head of the figure of the King that birds do not sit on it' (transl., Piers, 1916).

A number of running repairs continued to take place in the first half of the 14th century, but by the middle of the century, works had come to a halt. The castle continued to be maintained and garrisoned and was put to siege by Owain Glyndwr, who led a revolt against the English occupation of Wales in the early 15th century. Caernarfon Castle withstood siege during Glyndwr's rebellion twice. However, it seems that during the rest of the 15th century, the castle was used as a prison. In 1538, King Henry VIII ordered a survey of the Welsh castles, and the outcome was that all were in a state of disrepair and much overgrown. Five people were employed to cut the ivy from the walls, a job that took them a fortnight (Piers, 1916). A new stone chimney was built and general repairs were made. Accounts (again reported by Piers, 1916) tell that stone was brought from Maenan Abbey (which Henry VIII had dissolved) as well as the Lady Chapel and Friary at Bangor, other casualties of Henry's religious reforms. Piers (1916) points out that that it is not stated whether this material was specifically used at the castle. By the end of the 16th century, it seems that the castle was once more in use as a prison, but once again decay of the infrastructure, predominantly the timber and the entrance bridge, was significantly advanced to the extent that prisoners could all too easily escape its confines.

By the mid-17th century, Caernarfon Castle was initially garrisoned on behalf of the Royalist cause in the English Civil War but changed hands between various factions during this period. This was the last time Caernarfon Castle was used as a fortress, and like many castles garrisoned during the Civil War, it was ordered to be slighted. Luckily, however, it somehow managed to escape this fate, but sadly once again fell into disrepair over the next two centuries.

Caernarfon became a wealthy town in the 19th century with the meteoric growth of the slate industry, and the extension of the rail network provided a direct connection to London in 1852. Repairs and restoration, as described above, are more evident at Caernarfon than at the other three castles discussed below, and this culminated in a major campaign of restoration during the mid-19th century. However, in 1845, the castle was presenting a danger of collapsing onto the town below its walls. The mayor, Thomas Evans managed to engage the acclaimed architect, Anthony Salvin (1799–1881) to survey the building and make adequate repairs. Salvin was an expert on Medieval building and opted to conserve and consolidate the castle rather than to rebuild it. Thus began a series of works to bring the castle back into a presentable and safe state. The Constable of Caernarfon Castle from 1870 was Sir Llewellyn Turner (1823–1903), who initiated a major programme of renovation and rebuilding which he funded at least in part by charging an entrance fee to visitors. Turner employed a local mason, John Jones, to oversee the choice of stone and building works at this time. The investiture of the then Prince of Wales, the future Edward VIII was planned for 1911, and further improvements to the fabric were made in the run up to this event. The current Prince of Wales was also invested at Caernarfon in 1969, again requiring a few structural changes to be put in place. Avent (2010b) provides a lively account of the persons involved in the renovation of Caernarfon Castle during the 19th and early 20th centuries, including the work of the notorious mason John Jones, employed by Llewellyn Turner, who 'unchecked' in his work, rebuilt several sections of the Castle walls with granite, something that was to cause great annoyance amongst antiquarians and commissioners at the time. This lack of historical integrity was not forgotten, and when Edward Greenly studied the stone of the castle 60 years later in the in the early 1930s, Jones's anachronistic and cheap work still rankled with the then Secretary to the Royal Commission on Ancient and Historical Monuments for Wales, the archaeologist Wilfrid Hemp (Greenly, 1932). Nevertheless, Jones's skill as a mason was considerable. He was able to extract rotten blocks and replace them with such well-trimmed blocks of granite that even Greenly was fooled and he admits to initially believing the granite to be part of the original structure.

Caernarfon Castle consists of an Upper and Lower Ward (Figure 4.4), the former being the site of the Norman motte. These are enclosed by walls incorporating ten octagonal towers, including a double-turreted defended, gate house, the King's Gate, which opened across a drawbridge onto the town. This building features traceried windows and an imposing statue of the king above the entrance. A second gate house, also double-turreted, is

Figure 4.4 A view of the Interior of Caernarfon Castle, looking from the Queen's Gate west towards the Eagle Tower. The circular structure in the Upper Ward is the dais constructed for the investiture of the Prince of Wales in 1969. This is on the site of the former Norman motte. The foundations of the Great Hall can be seen on the left in the Lower Ward in the middle distance.

the Queen's Gate which would have opened onto a bridge and ramp which led down to the quay side at the eastern end of the Castle. Opposite this at the western end of the Castle is the Eagle Tower, so-called for the crumbling remains of stone eagles which would have adorned the merlions of the battlements. This is the tallest of the castle towers, and it is topped out with three octagonal turrets. A Water Gate was to be built here, to allow sea traffic to enter the castle directly, but his was never to be completed. The beginnings of a wall, containing a slot for a portcullis, remain *in situ* and resemble a buttress on the west side of the Eagle Tower. From the Eagle Tower, moving anticlockwise around the castle, towers are named as the Queen's Tower, the Chamberlain Tower, The Black Tower and then the small Cistern Tower. Then comes the Queen's Gate and the North East Tower, which connected with the eastern end of the Town Walls. Continuing around the circuit of the walls, the Granary Tower, King's Gate and Well Tower face the walled town, with the western end of the Town Walls connecting with the Eagle Tower. This final section of curtain wall, within the walled town and including the King's Gate, was the last section of the Castle to be completed. In the interior, a huge kitchen was built against the north wall, between the King's Gate and the Well Tower. Opposite the kitchen, the Great Hall was constructed against the south wall between the Queen's and Chamberlain Tower. The towers have three to four stories, forming spacious, octagonal rooms. The intervening curtain walls are some 6m thick and contain a circuit of intramural passages which allowed the defence of the seaward-facing south and west walls. A characteristic archway, wherein a horizontal lintel has been placed on top of a shouldered archway, is a typical architectural feature of Caernarfon Castle, and this style when seen elsewhere in British Medieval architecture has subsequently become known as the 'Caernarfon Arch' (Figure 4.5).

Caernarfon Castle is the only one of the four castles described here that does not utilise the bedrock it stands on as a building stone. It is built on an outcrop which straddles the boundary between the Padarn Tuffs, which underlie the walled town and the overlying siltstones and mudstones of the Nant Ffrancon Group. The latter are well exposed at the base of the south-facing curtain walls of the castle. To be fair, both of these lithologies are intractable as dimension stone, and apart from their use in dry stone walls and the occasional prehistoric monument, they have not been used for construction stone.

Figure 4.5 A Caernarfon Arch on the wall walks of Caernarfon Castle. The dressings are in Basement Beds gritstones and red Cheshire Sandstone.

Contemporary documents, predominantly Exchequer Accounts, as translated and published by Piers (1916) provide firsthand account of the procurement of stone for the building of Caernarfon Castle. Although stone from 'quarries' is mentioned in the documents pertaining to the early phases of building, their locations are not stated; however, many references to stone being brought in by boats suggest that quarries were situated on coastal outcrops on Anglesey and the mainland. There has always been an assumption that stone was taken from the Roman Fort at Segontium for the construction of the Castle and Town Walls. There is a long history of stone being robbed from pre-existing structures worldwide, so this is certainly a possible source of stone. Only foundations now remain for the most part of the excavated site of Segontium, suggesting that it had been substantially dismantled. Many buildings there may have only had stone foundations with timber-built upper stories. Nevertheless the fort, at the time the only substantial stone building in the region, had probably been seen as an easy source of stone since its abandonment in the late 4th or early 5th century CE. It may well have been the case that the castle builders initially relied on stone removed from the Roman fort of Segontium, with relatively minor amounts of stone brought in by boat from further afield (i.e. Anglesey). However, there is no reference to the spoliation of stone from Segontium in the Excheq-uer accounts, not that this should be interpreted to mean that stone was not taken from the Roman Fort. Locating quarries from his-torical sources is unfortunately infrequently an easy task. For the construction of Caernarfon Castle, named quarries occur in later documents relating to building works undertaken in the early 14th century which relate to the phase of repairs undertaken between 1315 and 1322, under the direction of Edward II and overseen by the mason Henry of Elreton. Although this is the first mention of the location of stone quarries in the documents relating to the con-struction of Caernarfon Castle, there is little reason to doubt that some if not all of these quarries had been used for the initial con-structions of the castle and town walls. Quarries on Anglesey are cited as the main suppliers of stone, but also Aberpwll, Mapbon, 'ponte meney' the 'black quarry' and a quarry 'ad finem ville' (at town end) are named. These locations are only tentatively located. *Mapbon* and the 'black quarry' have not been identified.

In addition to the potential of available sources of ready-cut masonry at Segontium, other buildings were also reported as being recycled into the castle, although again, evidence for stones from outside the main sources is not immediately obvious in the current structure. The Exchequer accounts (Piers, 1916) state that timber was brought from the buildings associated with the royal court at Aberffraw in 1317, which was itself located close to outcrops and quarries of good, freestone-quality sandstone at Malltraeth. The timber Hall of Llewelyn in Conwy was dismantled and brought to Caernarfon in its entirety, including a slate roof which was re-constructed in the castle by Henry la Sclatiere (Henry the Slater). Another building belonging to Llewelyn was reputedly taken to Harlech. The numerous mentions in the accounts of wood, either fresh-cut or brought from buildings which were dismantled show that a large number of structures within the castle bailey were timber-built with thatch or slate roofs.

We have also seen that religious buildings dissolved in the 16th century by Henry VIII were also dismantled and their stone brought to Caernarfon, including the abbey at Maenan, although there is scant evidence for stone from the Conwy Valley observed in Caernarfon Castle.

Major repairs in the 19th and early 20th centuries use new stones obviously out of keeping with the original masonry and sourced from outside the local area. Upper Carboniferous sandstone from north east Wales and granites from Ireland were used to replace staircases, window dressings and to shore up the Castle walls. A stone dais in Welsh Slate was constructed for the investiture of Prince Charles, the current Prince of Wales in 1969. The stones used in the construction of the castle and its subsequent repairs and renovations are reviewed below.

The main building stone of Caernarfon Castle is Lower Carboniferous, Clwyd Group Limestone. This stone is used in the exterior Castle walls, and here it is the predominant building stone, creating an impressive and uniform façade, especially in the walls facing the entrance to the River Seiont (see Figure 4.2a above). In the interior walls, the limestone is also used, but it is less abundant. It appears this high-quality stone was prioritised for the outer walls to provide an impressive appearance to the Castle, especially when viewed on approach from the sea. The use of limestone, cut into regular blocks and laid in coursed masonry, marks Caernarfon out from the other three Castles which, notably, do not use this limestone in their fabrics and do not use cut ashlar blocks. This is especially remarkable when considering Beaumaris Castle which is located only a few kilometres from the Penmon quarries and built almost a decade later than Caernarfon, Conwy and Harlech Castles, with all the experience this major building campaign would have afforded. Tradition has it that the limestone used at Caernarfon Castle was 'quarried' from the Roman Fort at Segontium, but there is little evidence to prove that this is the case, however, Segontium was a potential and plausible source of this stone. What is clear is that there is a very large quantity of this stone used in the walls and towers of Caernarfon Castle, possibly much more than would have been used at the fort, which may have only used stone for in the ground floors, with upper levels constructed from timber. It should also not be forgotten that Segontium had been abandoned for almost 1,000 years at the time of the building of Caernarfon Castle and had probably been raided for building materials throughout that time. However, Edward Greenly's argument that this high-quality building stone was used at Caernarfon (and only at Caernarfon), precisely because it could utilise *spolia* from the fort, carries some weight (Greenly, 1932).

The limestone used in the Castle walls is a fine-grained micrite; a compact, dark-grey limestone which develops a distinctive, pale-grey weathering crust. It contains scattered, detrital fossils;

particularly thin-shelled brachiopods, rugose corals and distinctive gastropods, around 1 cm diameter. These are sometimes infilled with calcite spar and weather out prominently from the surface of the stone. Many blocks are cut by parallel, calcite veinlets, which also weather out from the surface. All authors who have studied these stones (see Lott, 2010; Nichol, 2005; Greenly, 1932 and this study) have remarked that stone of this facies, quarried from the Cefn Mawr Limestone Formation, can be sourced from all outcrops of the Clwyd Group Limestones in North Wales; however, the Penmon area in Anglesey has been identified as the most likely source (Greenly, 1932). The dimensions of blocks are largely controlled by bedding thickness and the spacing of orthogonal joints present in these strata. Although well-cut and dressed, a range of block sizes are encountered, this strongly suggests that they were not originally cut for use in this building. Examining the façade of the castle, blocks of different sizes tend to be grouped together, say in a single course of masonry, suggesting that they were used as they were supplied and different block sizes did not have different functions. Such a pattern in masonry strongly suggests the reuse of blocks from previous structures.

A closer source of Clwyd Group Limestones is on the Vaynol Estate on the mainland shores of the Menai Straits, between Bangor and Y Felinheli. The Carboniferous limestone at Vaynol is distinctively iron-stained from overlying red beds and has developed a characteristic pink colour as a result. A few blocks of this stone do occur in the interior Castle walls and can be seen at eye level in the entrance to the north tower of the Queen's Gate and also associated with 19th century repairs in the Chamberlain Tower. This stone is also seen in the town walls. It may be the case that all examples of pink Vaynol Sandstone relate to 19th century renovation. The Vaynol limestone is from the Treborth Formation, which has now been reassigned to the Loggerheads Limestone Formation of the Clwyd Group.

Sandstones and gritstones, in the form of the Lower Carboniferous Basement Beds, are used in abundance in the construction of Caernarfon Castle. These sandstones both underly and are interbedded with the lower parts of the Clwyd Limestone Group throughout its outcrop on Anglesey and on the Vaynol Estate. A wide range of facies of clastic rocks are identified within these strata, ranging from medium-grained, buff-coloured freestones to coarse-grained gritstones and pebble conglomerates, which exhibit a wide range of colours varying from buff to green, purple, pink and red. Arguably the whole range of these sandstones is employed in the construction of Caernarfon Castle, suggesting a supply of stone from several quarries on Anglesey and on the mainland.

In the Exchequer Accounts relating to building works in the 14th century, individual quarries are named. According to these accounts, building stone came from quarries at 'Aberpwll'.[5] Aberpwll Stone was also used for paving on the walkway between the Chamberlain and Black Towers. Historically, Aberpwll was a village located in what is now the town of Y Felinheli (Port Dinorwic) which was later developed as a slate port in the 19th century. Geologically this region is dominated by Carboniferous limestones rather than Basement Beds sandstones. It is also speculated that it is possible that Aberpwll might refer to stone from Moel-y-Don on the opposite shore of the Straits, where a red sandstone from the Basement Beds outcropped. The peninsula of Moel-y-Don is now flat, but the Welsh toponym *moel* suggests a rounded hill. This may well have once been the case as this area has in fact been entirely quarried out with only small outcrops or red sandstone and gritstone visible around the margins of the present day fields and along the shoreline. Nevertheless a good idea of the nature of the Moel-y-Don stone can be discerned from the structures built on its outcrop (particularly the Elizabethan manor of Plas Coch and St Nidan's Church; see Chapter 5) and in drystone walls. The stone here varies from dark, liver-red sandstones to red-brown gritstones with laminations and strings of quartz pebbles. There are also the red and buff mottled, so-called 'peaches and cream' sandstones of Davies (2003). These red sandstones, which were also used at Segontium, have in the past been mistaken for the Triassic sandstones of Cheshire. Surprisingly, relatively little of this red sandstone is used in Caernarfon Castle. It is most frequently seen in the decorative stripes in the outer walls and towers. However, the sandstone in these stripes varies in colour from red to yellow, but with more yellow shades predominating. However, as the bulk of this stone has been quarried out, it is now impossible to reconstruct the

---

5   Block dimensions are, unusually, supplied for the Aberpwll Quarry stone consignment. Two sizes of blocks were supplied, dimensions are given as 2.5′ × 1.5′ × 1′ (76.2 × 45.72 × 30.5 cm) and 1.5′× 1.5′ × 1′ (45.72 × 45.72 × 30.5 cm) (Piers, 1916).

stratigraphy and facies within the Moel-y-Don outcrops of these fluvial sandstones.

Another named quarry is 'Ponte meney', literally 'Menai Bridge'. Clearly long pre-dating any permanent bridge across the Menai Straits, the 'ponte' was a pontoon bridge that had been constructed by Richard the Engineer during Edward I's 1277 campaign in North Wales (Prestwich, 2010), and this structure seems to have remained in place for some years. Assuming that the pontoon was at the narrowest part of the Straits, close to the location of the modern suspension bridge, this would indicate quarries on the mainland shore of the straits. These units outcropping in this region are known as the Menai Straits Formation (also the Bridges Sandstone) and include the Fanogle Sandstone Member on the Anglesey shore (Howells et al., 1985; Davies, 1982). These units form the base of the Carboniferous successions and underly the limestones. Greenly (1928) referred to these units as the 'Loam Breccias' which he describes as a series of poorly stratified beds of buff, red, green and purple gritstones and conglomerates with beds of clay-rich, shaley sands (the 'loams'). Certainly stones matching these facies are widely used in the interior of Caernarfon Castle. Green-coloured, trough cross-bedded sandstones are often seen as window dressings in the interior walls.

Wherever the specific source, these Basement Bed gritstones are the main stones used in the Castle interior for masonry, paving, dressings and quoins, merlions and other architectural features such as water-spouts, corbels and even decorative carving. The poor state of the stone eagles on the Eagle Tower is directly attributable to the somewhat inappropriate stone used to shape them. On the exterior gritstones are used for the decorative stripes on the walls as well as dressings on the windows, arrow loops and doorways. It is also used secondary to the limestone in the facings of the exterior walls and more commonly in the lower half of these walls than in the upper. This is also a building stone used in significant amounts at Segontium, where there is extant evidence that this stone was well dressed. It is plausible that at least some of the Basement Beds sandstones and gritstones from Segontium were incorporated into the Castle.

The stripes on the walls facing the River Seiont are grouped by colour. There are three main stripes on the south face of the castle, the lower one created by two courses of sandstone masonry, and the upper two have a thickness of three courses in the otherwise limestone-built façade. The lower two stripes are of predominantly of iron-stained, yellow and buff-coloured Basement Beds sandstones. The uppermost stripe is made almost entirely from green-grey coloured, well cross-bedded gritty sandstone which is clearly derived from a different quarry source, possibly from the Bridges Formation strata. This uppermost band also incorporates a series of arrow loops which are dressed in the same stone.

A coarse-grained sandstone bed forms the base of the Ordovician Nant Ffrancon Group and directly overlies the eroded upper surface of the Twthill Granite in Caernarfon. A prominent marker horizon, this unit, officially named as the Allt Llwyd Formation varies in thickness along strike up to a maximum of some 65 m where exposed at Crûg, some 2–3 km NE of Caernarfon. In Caernarfon, the full thickness is not exposed, but it is probably no more than 10 m thick. It is a medium-grained, arenitic metasandstone, with dark 'films and knots' of carbonaceous material. In colour it appears blue-grey, and it is laminated on various scales. When observed in buildings, this stone is in the form of squared-off blocks, cut parallel to bedding and with block sizes equivalent to the thickness of the bedding planes, usually about 10–15 cm thick and 20–30 cm in length. It is probably a very hard and difficult stone to work, but the work of the masons would have been much aided by the strong parting along bedding planes. The stone is used for coursed masonry in the town-facing wall of Caernarfon Castle. This sandstone had been remarked upon by Edward Greenly, and he identified it as the gritstones overlying the Twthill Granite which he had at the time assigned to Cambrian Strata. Quarried ground, much overgrown and now in private ownership certainly exists on the north end of the Twthill, and Greenly interpreted this location as the 'Town End Quarry', referred to as the quarry 'ad finem ville' in the Exchequer accounts (Piers, 1916). Although this quarry had been expanded in the 19th century for the extraction of granite, Greenly had found traces of what is now known as the Allt Llwyd Formation sandstones in the quarry and tentatively identified it as the source of the stone.

Outcrops of this stone in the town of Caernarfon have been more recently revealed in a road cutting of the A487 road which bypasses the main town centre. This road and its junction with local roads at the base of Twthill were constructed in the 1970s. The exposure of the Allt Llwyd Grits here, on the SW side of Twthill, may well not have been known by Greenly but confirms his understanding of the stratigraphic location of these beds. This sandstone is of importance as it is not hitherto well recognised as a building stone, and it is also the only obviously locally quarried stone used in Caernarfon Castle. It is used in the early 14th century north, interior wall of Caernarfon Castle and is easily examined at eye level in the kitchen and shooting gallery, both located in the north-facing, town-facing wall overlooking Palace Street (Figure 4.6). It is also used in 19th century repairs on the Chamberlain Tower (Figure 4.7).

Imported stone from NE Wales, England and beyond is also encountered in Caernarfon Castle. High-quality, red sandstones of Middle Triassic (Keuper) age are quarried in Cheshire, and specifically the Delamere Sandstone Member of the Helsby Sandstone Formation was used, albeit in limited amounts, at Caernarfon Castle. Its bright red colour contrasts starkly with the buff and grey building stones of the castle. These are well-laminated and cross-bedded, red sandstone with medium to coarse grain size (so-called 'millet seed' sand grains) and a few strings of quartz pebbles, up to 2.5 cm diameter. Pebble-free beds also occur and are more abundant. The stone is variably micaceous.

Figure 4.6 The kitchen area in the interior of the Lower Ward of Caernarfon Castle. The storage niches are constructed of uprights of Basement Beds gritstone. Allt Llwyd Grits are used as coursed masonry forming the soffits of the niches and the wall above.

19th Century restoration:
Vaynol Limestone

19th Century restoration:
Penmon Limestone

19th Century restoration:
Allt Llwyd Sandstone

19th Century window dressings of
Gwespyr Sandstone

Original masonry (late 13th Century) of
Basement Beds Gritstones and Cefn Mawr
Limestone

Figure 4.7 Restored windows and masonry in the Chamberlain Tower of Caernarfon Castle. Gwespyr Sandstone has been used for the window dressings and the overlying masonry has been patched with Allt Llwyd Sandstone as well as Penmon and pink Vaynol Limestone.

In the 1390s, building accounts show that stone was bought from Chester to Caernarfon for repairs to the 'King's Tower of the Banner' (La Tourebanere), now known as the Queen's Tower (see Figure 4.5). These works were carried out by one William of Frodshame,[6] who was then the Chamberlain of North Wales and presumably a Cheshire man by origin (Piers, 1916).[7] This stone is also used on the Queen's Gate, both as window and door dressings and for repairs to the spiral staircase in the tower. Strings of pebbles can be seen on the surfaces exposed in the tower stairwell.

This documentary source secures the provenance of the red Triassic sandstones used at Caernarfon Castle, and the evidence seems to indicate that this stone was brought directly from Cheshire for this specific phase of repairs, organised by a man who would have known the Cheshire stone well and may even have had stakes in the quarries themselves. However, historically both Edward Greenly and the archaeologist who had excavated Caernarfon's Roman Fort at Segontium believed that the stone had been taken from the site of the fort. Greenly (1932) notes that Cheshire sandstones were not employed at Beaumaris, despite this being somewhat nearer to the sources. His reasoning for its absence at Beaumaris is indeed that Beaumaris was not close enough to Segontium. The stone used at Segontium was identified as the Delamere Member of the Helsby Sandstone Formation by Sir Aubrey Strahan (Wheeler, 1924); however, further examination suggests that these red sandstones were in fact sourced much more locally, from the Moel-y-Don Basement Beds of Anglesey. Although this myth has perpetrated into some recent literature, there appears to be scant evidence of Triassic Cheshire sandstones being employed at Segontium, and therefore this could not have been the source of Cheshire sandstones observed in Caernarfon Castle.

Running repairs to Caernarfon Castle have occurred throughout its history. There has been much more restoration and maintenance at this site than at the other castles, and non-local stones have been used for these works. Following the dissolution of Maenan Abbey in the Conwy Valley in 1538, stone and timber were taken

---

6 Frodsham is a village in Cheshire located on the escarpment formed by these Triassic sandstones.

7 The Exchequer accounts also report that timber was felled in Delamere Forest and Shotwick Park for building works at Caernarfon Castle at this time.

and used for building at the nearby Gwydir Castle and also reputedly taken to Caernarfon for repairs to the castle and town walls. The same arrow-shaped mason's mark has reputedly been found on red sandstone recovered from excavations at Maenan Abbey, and similar marks have been found on stone (undefined) at both Conwy and Caernarfon Castles (Owen, 1917; Butler & Evans, 1980). However, no lithologies from the Conwy Valley region have been observed in Caernarfon Castle.

Repair and rebuilding work carried out in the 19th century is far better documented, and in the early 20th century, Edward Greenly was still able to talk with people who had continuity with work carried out in the mid-19th century. Some stone is recorded and discussed below, but several areas of patching using Allt Llwyd gritstone and pink Vaynol limestone are frequently seen with the fabric. Repairs made to the Queen's Gate of the Castle in the 1840s under the architect Anthony Salvin (1799–1881) 'utilised a brownish sandstone from Talacre' to reconstruct the treads of the spiral staircases (Piers, 1916). This is Gwespyr Sandstone, a medium to fine-grained arkosic sandstone, composed of quartz, feldspar and variable amounts of mica. It is bedded and cross-bedded on a variety of scales. Colour varies from buff through grey to, in places, a pink-red colour. Howe (1910) describes it as having a distinct greenish tint, and this appears to be the most common facies of the stone observed generally and particularly at Caernarfon Castle. Tafonic weathering, forming cavities defined by bedding and cross-bedding structures, is common in older uses of this stone, as are iron spots and Liesegang banding. These latter features can be clearly seen on exposed examples of Gwespyr Sandstone used as replacement window dressings in several locations in the Castle.

The Gwespyr Sandstone Formation belongs to the Upper Carboniferous Millstone Grit Group. There is a fairly limited outcrop and sub-crop of this sandstone at Talacre where the principle quarry was situated. The same strata were also worked at Berwig near Wrexham. As one North Wales's most prominent building stones of the modern period, Gwespyr Stone was also used in the later 19th century repairs to the Caernarfon Castle conducted by Sir Llewellyn Turner and his mason John Jones. Jones used this stone to replace window mullions and arrow loop dressings. Unfortunately, this stone was not quarry laid and now under attack from weathering, presents a rather piecemeal set of repairs, as can be seen in the window dressings in the Chamberlain Tower (Figure 4.7).

Despite his clear skills as a mason and builder, Jones's work was much despised. Sir Schomberg McDonnell, secretary to the Office of Works, wrote in 1911, 'There was nobody to say nay or to control [John Jones], and it is a fact that at this moment every one of the new battlements which deface [Caernarfon Castle] were constructed, not out of local stone, but out of York stone [sic] specially procured for the purpose' (quoted in Avent, 2010b). McDonnell was mistaken in the identity of the stone, he was referring to Jones's use of Gwespyr Stone.

It is also possible that the early 14th century statue of King Edward on the King's Gate is also Gwespyr Sandstone. If so it is a very early use of this stone, but its fine grain size would have rendered it more attractive as a freestone and therefore more fitting for a royal effigy than the coarse-grained local sandstones and gritstones. Much weathered and the iron cramps that hold it in place still clearly visible, the statue is cut from a cross-bedded medium to fine-grained sandstone. It is too high up to identify the stone with any certainty. Bedding planes are vertical on most of the blocks used. Weathering has revealed clear cross-sets. The base of the bust is on a horizontal slab of fine, greenish-coloured buff sandstone with tafonic weathering. These features and properties indicate that the statue is probably made from Gwespyr stone rather than more local Lower Carboniferous sandstone.

A pale grey-coloured granite, or in fact granodiorite, is associated with Sir Llewelyn Turner's restorations of the 1870s and is used on the seaward walls and battlements. The stone is neatly and, at first glance, seamlessly, inserted into the original masonry of the walls, sometimes precisely replacing a single block. This stone is not local. It was brought into the slate ports of Caernarfon and Port Dinorwic (Y Felinheli) as ballast in large quantities and is widely used in both towns for construction of 19th century houses, civic buildings and churches. Once again, its use in the castle was the choice of the mason John Jones. Edward Greenly was informed by the archaeologist Wilfrid Hemp that Jones was able to buy the ballast very cheaply and overlooked all other stone as a consequence, despite the granite being unsympathetic to the fabric of the building. The stone is a medium-grained granodiorite, composed of white plagioclase feldspar, black hornblende and biotite and grey quartz. It is a good match in appearance for two stones which were actively quarried at this time; either Newry Granodiorite from the Republic of Ireland or alternatively Kirkmabreck

Granite from Dumfrieshire in Scotland (Greenly, 1932; Nichol, 2005). These two stones are indistinguishable in hand specimen. Greenly (1932) believed this stone to be from Ireland and was brought into Caernarfon as ballast from the slate ships, whereas Nichol (2005) somewhat more implausibly suggested that it was Kirkmabreck Granite on the basis that this stone had been used for construction of the Liverpool Docks. Both authors overlooked the preponderance of this stone used in other buildings in the slate ports but not elsewhere in the hinterland of NW Wales. It common occurrence in 19th century vernacular building strongly supporting the view that this is ballast, and as told to Greenly by Wilfrid Hemp, derived from the slate ships retuning from Ireland. Caernarfon, particularly, had had a century-long tradition of supplying Dublin and Belfast with roofing slates. The author has studied thin sections of this stone collected from building renovations in Y Felinheli. Both the Newry and Kirkmabreck Granites have very similar mineralogy and petrology, but the presence of abundant micrographic-intergrowths between quartz and feldspar, which is rare in Kirkmabreck Granite, indicates that indeed, Newry is the most likely source of this stone (Figure 4.8).

4.3.2   Caernarfon town walls

The town walls of Caernarfon form an almost complete circuit enclosing the town. There are ten towers in the walls which are numbered from the east, anticlockwise, 1–10 (Taylor, 1993), and they will be referenced using this numerical marker below. A lot less care in the selection of uniform stone was employed in the construction of the walls than is the case in the Castle and, notably, the towers are circular in plan rather than octagonal. Also the walls do not have the decorative stripes observed in the Castle. The walls are for the most part constructed of coursed masonry. The majority of the stone blocks used have been squared off, though a number of boulders and glacial erratics have also been incorporated into the fabric.

The Queen's Gate at the eastern end of the Castle overlooks the marketplace, Y Maes. This was once the town green, and beyond this lay the artificial fish pool which drained into the River Cadnant which ran down to the Menai Straits alongside the eastern town walls. The River has now been culverted and now runs beneath

Figure 4.8 (a) Caernarfon Castle Foundations on the Nant Ffrancon Formation bedrock. The blocks used here are of predominantly granodiorite used in the repairs undertaken by mason John Jones in the 1870s. This stone is also widely used for walls and other buildings constructed in Caernarfon and Y Felinheli in the mid-19th century. (b) A sample of the same granite, collected from garden rubble, in thin section. The granodiorite is composed of potassic feldspar, plagioclase, hornblende and biotite, with large, euhedral crystals of sphene and allanite.

Crown Street (Glan Ucha) and Bank Quay in the modern town and into the SE corner of the harbour, Doc Fictoria.[8] The Cadnant was an important working river in the medieval townscape, providing not only water for the fish pool but also fuelling mills and wharves along its course.

The town walls connected to the Castle, north of the Queen's Gate at the North East Tower. This first part of the walls, which crosses the Castle Ditch, is the only section now missing from an otherwise complete circuit. There would once have been a small postern here, the Green Gate (Taylor, 1993). This break in the wall at the corner of Y Maes (Castle Square) with Stryd Twll-yn-yr-Wal (Hole-in-the-Wall Street) allows an exceptional section of the foundations of the wall's core and therefore its construction to be observed. The wall can be seen to be built with a rubble core which is faced with stone. The interior rubble is composed of large, rounded glacial cobbles, between 20 and 50 cm diameter. These stones would have been collected from nearby glacial tills which outcrop adjacent to the River Seiont and in the hinterland of the town. They are predominantly volcanic lithologies from the Ordovician igneous rocks of Snowdonia. These include a prominent white-weathering boulder of a rhyolitic tuff with numerous, 2 cm diameter lithophysae (volcanic gas bubble) which have been infilled with quartz. Numerous other blocks of acidic tuffs are present as well as porphyritic andesites and dolerites. These stones are set in a lime cement with a coarse aggregate; the mortars visible today are evidence of recent repointing. This section through the wall shows these boulders to be roughly coursed and laid in successive layers which were then left to dry before the next was added. The lowest course seen at this location is made of more uniformly sized (c. 20 cm diameter) cobbles. The layer above this has variably sized boulders. The eastern side of the wall is obscured by an early 20th century wall in Penmon Limestone and a plaque in the same stone commemorating the gift of this section of the walls to the town by local landowner Owen Jones in 1917.

The circuit of walls extending along the eastern side of the walled town along what is now Greengate Street were destroyed during the Welsh uprising of 1294 and subsequently repaired.

---

8 Prior to the development of Crown Street, now Glan Ucha, in the 20th century, this route was that of the railway which came into the Slate Quay.

A large amount of pink-stained limestone from the Treborth Formation of the Clwyd Limestone Group and quarried from the
Vaynol Estate has been used in this section of the walls (between
towers 1 and 3). Unfortunately, this is well above eye level and difficult to observe at close quarter. Tower 3 is the double, D-plan towers of the main gate into the town, the Porth Mawr or the Exchequer
Gate. This is where tolls and taxes were administered and was the
main entrance to the walled town. The road to this gate would have
crossed the River Cadnant using a drawbridge and access is still
across a modern bridge. This gate way has been much modified,
first in 1767 when it became Caernarfon's town hall and again in
1833 and 1873 (Taylor, 1993). The gate as its stands today is mainly
constructed from Penmon Limestone (including on the vertically
extended, upper portions of the towers) and some Basement Beds
gritstone. The latter mainly found in the upper courses of the interior sides. A plaque in the interior of the gatehouse arch, in spotted
blue Cambrian Slate (Llanberis Formation), commemorates the
renovations and restorations of 1833, including the installation of
the clock, which has a face on each side of the gate. The outer gate
entrance is dressed with cross-bedded Cefn Sandstone imported
from north east Wales and further embellished with polished pink
granite columns of Scottish Peterhead Granite from Aberdeenshire. These elements were put in place during the 1873 renovations.

The sections of the walls between Towers 5 and 6 are largely
original and are almost preserved to battlements level. The remains
of crenulated battlements and the lower parts of arrow loops are
visible. The arched entrances which provide access to Stryd Pedwar
y Chwech, Market Street and Church Street are relatively recent
additions (18th to 19th centuries). The Chantry Chapel of St Mary's
Church was built into the NW corner of the walls in the earliest
14th century. The interior arcading of the Church is in Basement
Beds gritstones.

The walls along the banks of the River Seiont, between St
Mary's Church Tower (Tower 7), and the Castle are the most accessible and it is here that the use of stone is most easily observed
(Figure 4.9). This section of walls was originally a wooden palisade
which was burned during the 1294 uprising and then reconstructed
in stone subsequently. The double-gabled west end of St Mary's
Church is incorporated into the walls just to the south of Tower 7
and a small door, originally a postern gate is blocked off, but still
visible within the wall work. Tower 8 has been much modified in the

Figure 4.9 Caernarfon town walls and Promenade. Tower 7 can just be seen behind Tower 8. The Promenade is built on the site of the Medieval wharf.

19th century and converted into a private residence. The windows, battlements, gothic-style doorway and chimney date from this time.

The West Gate, also known as the Golden Gate (*Porth-yr-Aur*), has also been much modified in 1870 when it was converted into the Royal Welsh Yacht Club by Sir Llewellyn Turner. An upper storey, battlements and windows were added at this time. Like the Great Gate, which lies opposite this gate to the west, there are twin D-plan towers. This gate would have been the main entrance from the quays and could once be closed by a portcullis. The final gate in the wall circuit, the Water Gate is located at the west end of Castle Ditch between Tower 10 of the town walls and the Eagle Tower of the Castle. Again this exit from the town is relatively modern (probably 18th century). The original plan early 14th century plan, which was started but never finished, was to have a huge water gate located here with a large portcullis. Traces of this structure, a buttress with a vertical slot to hold the portcullis, can be seen projecting from the Eagle Tower, a few metres to the west of the join with the current wall.

Approximately equal amounts of Basement Beds gritstones and Clwyd Group Limestones are used in the walls. The limestones are mainly of the same facies that is used in Caernarfon Castle, a dark-grey, fossiliferous micrite which weathers a pale grey. This is a facies of the Loggerheads Limestone quarried at Penmon. Blocks of reefal facies of this stone are also used, with well-preserved fossils of rugose and colonial corals, though these stones are less abundant in the fabric of the wall than the gritstones. Also present are the more shaley, rusty weathering, laminated, dark grey facies of the Leete Limestone which can contain large productid brachiopods. A few limestone blocks with calcite-infilled syneresis cracks (formed by subaqueous shrinkage of sediment) occur, as well as blocks of brecciated limestone. Such features are associated with palaeokarstic horizons with the Clwyd Limestone Group (Davies, 1983).

A wide variety of facies of Basement Beds gritstones are used. Most are yellow to buff weathering, with prominent beds and cross-beds of conglomerates. The main clasts types observed are of vein quartz, but clasts of red, Anglesey-sourced jasper are not uncommon. A few blocks of lavender-purple coloured Basement Beds grits, with vertical trace fossils typical of the so-called Loam Breccias of the Bridges Sandstone (Menai Straits Formation). Other blocks are of the Moel-y-Don and Pwllfanogl red varieties of these gritstones from the Anglesey shore of the Menai Straits. Some of these facies are fine-grained and show evidence of syn-sedimentary deformation.

A few blocks of a distinctive, very weathered, chlorite-rich greenschist with large quartz augen are observed in the section of the Town Walls around St Mary's Tower (Tower 7). Greenly (1928) describes boulders of these rocks, up to 30 cm diameter as occurring within his 'Loam Breccias' of the Lower Carboniferous Menai Straits Formation. Greenly had of course defined and mapped the Mona Complex of Anglesey and immediately recognised these distinctive augen schists as being from the Penmynydd Terrane outcropping both on the island and also the small group of islets called The Swellies in the Menai Straits. It is very likely that the augen schist blocks seen in the walls were 'quarried' as ready-made boulders from the Basement Beds Grits worked in the 'ponte meney' quarry areas rather than having been collected directly from the Mona Complex strata of Anglesey.

A few glacial boulders, probably sourced from local boulder clays and picked up from the beaches, occur in the fabric of the walls, though these are not particularly common. Some large examples of such boulders can be seen in the foundations of the walls between Towers 8 and 9. These are ~40 cm in diameter. A prominent example is a block of dark green, well-foliated amphibolite gneiss from the Neoproterozoic strata of Anglesey, more specifically the Coedana Gneiss Complex (Figure 4.10 and Horák, 1993).

The Royal Welsh Yacht Club occupies the West Gatehouse (*Porth-yr-Aur*). Conversion of this building into the Yacht Club's premises took place in the 19th century, and the structure of the building was much consolidated during this time. The main and most striking addition in terms of building stone are the window dressings in striking, red sandstone. Otherwise the main stone used is once again Basement Beds gritstones, Loggerheads Limestone and a few blocks of shaley Leete Limestone. The red sandstone exhibits well-developed bedding and cross-bedding laminations, as well as examples with syn-sedimentary deformation or dewatering slump structures. The stone used here is the Helsby Sandstone Formation of the Sherwood Sandstone Group, a Triassic sandstone quarried near Chester. The Yacht Club windows have been carefully made to mimic those in the architecture

Figure 4.10 The author standing next to Caernarfon town walls in the section between Tower's 8 and 9. The walls are constructed of a mixture of limestone and gritstone coursed rubble masonry, with random blocks of glacial material. A large boulder of Coedana Gneiss can be seen just left of centre. (Photograph by Jane Siddall.)

of the Castle, with a pointed, gothic window modelled on those in the Great Gate and tripartite, mullioned windows with miniature Caernarfon arches.

The soffits of the Water Gate are constructed of squared ashlar blocks of Clwyd Limestone mixed a few blocks of Basement Beds gritstone. Some of these limestone blocks, which would have been broken up along orthogonal joint surfaces, show evidence of copper mineralisation, with thin encrustations of green and blue copper carbonates (malachite, azurite and other copper salts). Large-scale copper mineralisation occurs within the Clwyd Limestones on the Great Orme peninsula (near Llandudno) and also within Ordovician rhyolites near Amlwch in north Anglesey (Parys Mountain) and elsewhere in the region. It is not impossible that small-scale pockets of copper-rich fluids had percolated the quarried Clwyd Limestones outcrops; however, evidence of copper mineralisation is rarely encountered in the building stones. There does not appear to be any evidence of metal fixings here which could have produced copper corrosion products, and texturally, these encrustations resemble copper mineralisation rather than staining from corrosion.

During the 14th century, the stretch of town walls facing the mouth of the Aber Seiont (between Towers 7 and 10 and the Eagle Tower of the Castle) was the location of the main port of the town. Here there had been an intention to build the large water gate which was never completed and certainly it would have been a good location for the Castle to exact tolls on imports and exports. Some of the original quay, used primarily for bringing materials for the construction of the Castle in the last two decades of the 13th century, was destroyed in the 1294 uprising. This was probably an earth and timber structure (earth and rubble in large quantities were being excavated from the Castle Ditch). Records indicate that this quay was patched up in the next 5 years and then rebuilt in stone after 1316. Stone from Aberpwll was ordered in 1322 (Cadw, 2010). Aberpwll was the settlement on the mainland shores of the Menai Straits (Y Felinheli), the local stone was Bridges Formation Basement Beds gritstones. The location of the intended water gate was superseded by a slipway, presumably built of the same stone. Further repairs were carried out in during the 16th century reformation, when stone was reportedly brought from the dissolved Friary at Bangor in 1538 (Cadw, 2010). This too is likely to have been Basement Beds gritstone.

4.4 HARLECH CASTLE

In terms of a defensive site, none of Edward I's castles is more prominent than Harlech Castle (Figure 4.11). Approaching the Castle and town across the flat coastal plain from Porthmadog, or indeed descending the passes through the mountains to the east, the castle can be seen for miles. It is hard to imagine that such a commanding site was not occupied prior to the wars and the subsequent invasion, but there is scant evidence in support of this. The Welsh folklore epic 'The Mabinogion' alludes to the rock of Harlech as the court of the giant Bendigeidfran (Davies, 2018), but there are no sound documentary records suggesting the presence of a *llys* or court belonging to the Welsh Princes (as at Criccieth) at Harlech, let alone one belonging to a giant. Similarly, one can easily envision a rocky, Iron Age hillfort to have been located in this rocky hilltop, although again no structural evidence for such a site has been found in excavations. A Bronze Age gold torc and some Roman coins are the only material cultures that indicate any previous occupation found on the site (Ashbee, 2017a); there is nothing that suggests that there was anything more permanent than the occasional camp. At the time of the founding of the castle, the local population was small with only twenty-three adults and twenty-one children resident in the village (Cadw, 2018). Although the population would have grown with the construction of the castle, Harlech has never been

Figure 4.11 Harlech Castle.

a large town, mainly due to its isolated position regarding trade routes or industrial centres.

Construction of the Medieval castle, under the supervision of Master James of St George, began in 1283 wherein the walls of the outer ward, much of the inner ward including the gatehouse complex, south-east and north-east towers were built. By 1284, the building was substantial enough to be garrisoned. In 1289, the south-west and north-west towers were built, and the west-facing wall of the inner ward was completed. The buildings of the inner ward also including the Great Hall, kitchens, storerooms and chapel were also completed during this building phase. Enclosure of the Castle Rock was initiated at this time, but this wall, of which little remains today, was not finished until 1295. Little subsequent building works have taken place. Some repairs or restoration of the Ystumgwern Hall on the south side of the Inner Ward was made in 1305–1306, and a barracks was built during the English Civil War in the 17th century against the north wall of the Inner Ward. The names of features of the castle are those used in the plan in Figure 4.12.

In terms of its use of stone, Harlech castle is probably the simplest to describe. It is built uniformly of Cambrian sandstones from the Rhinog Formation of the Harlech Grits Group, the bedrock on which its stands. The greatest geological interest lies in its supremely defensive site. The Castle is built on a bluff of Rhinog Formation sandstones and gritstones which on the west side rises almost vertically from the coastal plain, the Morfa, to a height of just under 60 m. At the time of building, the Morfa may have been at least partially flooded and was marshland.[9] The location of the shoreline is not well understood but access to the castle by boat to the foot of the Castle Rock seems to have been possible. A small jetty, the Water Gate, was constructed to allow access from the sea. This dramatic break in slope is controlled by the footwalls of the Glyn Valley and Mochras Faults which intersect at the point of the Castle. The Mochras Fault is a predominantly N-S to NNE-SSW trending structure which juxtaposes Cambrian strata of the Harlech Dome against Tertiary fluvial deposits (Allen & Jackson, 1985a). This structure is clear as a topographic feature and on the BGS map, and it appears to have a throw of between 4 and 6 km,

---

9 The Morfa was drained during land enclosure in 1806 (Cadw, 2018).

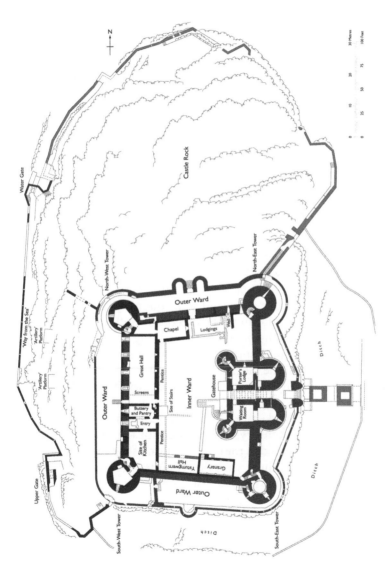

Figure 4.12 A plan of Harlech Castle. (Crown copyright Cadw (2020).)

and the structure is visible in outcrop for some 17 km between the Mawddach Estuary and Tremadoc (Figure 4.13). The main phase of fault movement probably took place during the later Oligocene. A borehole was drilled into the hanging wall at Mochras Farm in 1969 revealing a 1,300 m thick sequence of uniform mudstones of Early Jurassic age, overlain by Tertiary (Oligocene) fluvial sediments (see Allen & Jackson, 1985a; Hesselbo et al., 2013). The Mochras Borehole has provided the thickest succession of strata of this age in the United Kingdom and has potential to reveal important information regarding climate change and evolution during this complex period of Earth history.

The sandstone from the Rhinog Formation of the Mawddach Group outcrops in the cliffs beneath and around the castle and can be seen *in situ* in outcrop around the entire circuit of the castle and of course on the area known as Castle Rock to the north of the building. Outcrop is most easily accessed on the west terrace, where beds of Rhinog Formation dip gently towards the north-east. However, prior to construction of the Castle, this hilltop would have been a rocky and quite probably a rock-strewn space and some considerable quarrying would have needed to be undertaken to level the site, dig foundations and generally make it good for building. This in turn would have produced some quantity of building stone of various quality and dimensions. Blocks would have been relatively easily extracted by breaking them along bedding planes, and bedding thickness and jointing would have a strong control of the dimensions of the stone blocks used. Further materials could have been obtained locally from the area of the township itself as well as from quarrying the castle ditch, an exercise which began in 1285. A line of quarries worked into the seaward-facing cliff were in use until the early 20th century. Various qualities of quarried Rhinog gritstone have been used in the construction of the castle, from fine-grained to coarse-grained varieties (Figure 4.14). Lott (2010) has conducted petrographic analyses of both the green gritstones used in the castle and those of local outcrops and found them to be the same, quartz-rich greywackes in a clay-mica matrix.

Egryn Stone is a name applied by architectural historians to distinctive sandstone freestone used in a number of medieval constructions in the southern part of Gwynedd and throughout the county of Merionethshire including at Harlech Castle. However, stratigraphically this stone is poorly understood and quarry sites appear to be long worked out. Palmer (2007) has identified Egryn

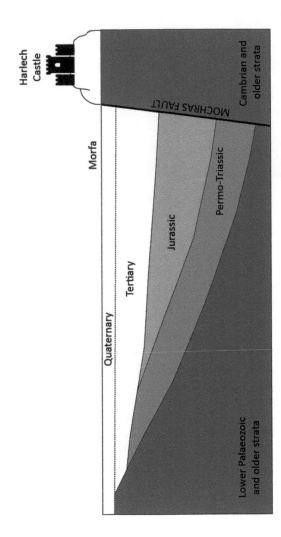

Figure 4.13 An interpreted section across the Mochras Fault. (Adapted from data in Hesselbo et al. (2013).)

Figure 4.14 Rhinog Formation sandstones and gritstones used in the construction of Harlech Castle.

Stone in over twenty medieval buildings (castles, abbeys, churches and houses) throughout this county, and it has also been identified by the author at St Beuno's Church in Clynnog-Fawr (see Chapter 5). This stone is distinctive, varying in colour from buff and yellow iron-stained to rarer blue-grey varieties. Iron staining may also occur in small spots across the stone and as Liesegang banding. It is a medium to coarse-grained, quartz-rich sandstone, exhibiting a weak cleavage and showing some elongation of the quartz grains parallel to this cleavage, and it is cross-cut by quartz veins. Documents describe a stone from 'Egrin' in use at Harlech Castle. Egryn is a manor farm near Barmouth, 15 km south of Harlech, and the quarry on the land is in the Llanbedr Slate Formation. Nevertheless, Palmer (2007) has identified that there are traces of quarrying above the main slate quarry excavating a narrow bed of buff-coloured sandstones, i.e. the Egryn Stone, which appears to be a unit transitional with the overlying green-coloured Rhinog Sandstone Formation. This unit has not been differentiated by the British Geological Survey. It may have been worked from other small quarries sites along strike, but it is no longer obvious in outcrop. Lott (2010) has analysed samples of the stonework attributed to Egryn Stone at Harlech Castle and found it to be texturally and petrographically very similar to that of the Rhinog Sandstones;

quartz grains supported in a matrix of mica and clay. They are greywackes (quartz-wacke sandstones, Lott, 2010).

Egryn Stone is used at Harlech Castle primarily as dressings for windows and doors (though squared off blocks also occur in the coursed masonry of the walls). Its quality as a freestone makes it more amenable to carving decorative mouldings, though arguably such decorative features at Harlech Castle are somewhat austere. The yellow-brown to buff-coloured weathered surface makes it distinct from the grey-green Rhinog Stone, and it is easily discerned framing the windows in the gatehouse and doorways from the west wall of the inner ward accessing the outer ward (Figure 4.15). Building records indicate that these dressings, along with the corbels of the gatehouse turrets, were installed at the end of the main building phase, completed in 1289.

Few other stones are observed in the fabric of Harlech Castle. Slatey lithologies, derived from the Llanbedr Slate Formation, are used sparsely, and these strata outcrop locally, beneath the Rhinog Formation at the base of the cliff and of course abundantly elsewhere in the region, including as discussed above at the Egryn quarries. Rounded boulders and cobbles, ultimately derived from glacial deposits but probably collected from the local beach below the Castle, are also used, though they are uncommon. These are most frequently seen incorporated with Rhinog Sandstone rubble in wall cores. Geologically, these are rounded boulders of Rhinog Formation sandstones and cobbles of white vein quartz.

Figure 4.15 A west-facing doorway in the Great Hall of Harlech Castle, dressed with Egryn Stone.

4.5   CONWY CASTLE AND TOWN WALLS

4.5.1   Conwy Castle

Conwy Castle stands on an imposing natural site, situated on the west side of the estuary of the navigable River Conwy, the major river of the region. It is located on a rocky overlook, surrounded by steep cliffs, which on the south side drop down to the mouth of the River Gyffin (Afon Gyffin) where this flows into the River Conwy. The castle thus commands both sea and river trade as well as the crossing point of the River Conwy, once by a ferry and now by bridges constructed in the 19th and 20th centuries. The adjacent town is enclosed by a complete circuit of walls and still retains much of its original street plan (with only minor 19th century redevelopment). Conwy Castle and the circuit of town walls are extremely well preserved and have required little in the way of rebuilding since the late 13th century (Figure 4.16).

An important and strategic conquest in the campaigns of 1282–1283, Edward I had taken Conwy in early Spring 1283 and by May of that year, quarrying had already begun on the castle rock and the town had been enclosed by a wooden stockade. At this time Edward himself had been resident in the town (Ashbee, 2015). The site was certainly attractive as one of offensive command and defence, and the castle and new town were built on the site of an important local religious centre, the Cistercian Abbey of Aberconwy established in 1192. The Abbey had substantial regional significance as the burying ground for the Welsh Princes, and there was also a royal *Llys* or court (Llewelyn's Hall) in the town. Edward's first act was to demolish the Abbey and have it removed upriver to Maenan where it was rebuilt by Master James of St George. Whether or not the stone from the dismantled Abbey (if indeed it was built from stone) was transported upstream for reuse or was salvaged to be used in the Castle is not known. Llewelyn's Hall, probably a timber-built structure, was also dismantled and re-erected in Caernarfon Castle in its entirety, including its slate roof (Piers, 1916).

The Abbey Church of St Mary was retained as the parish church and is still situated at the centre of the town. The Castle and Town Walls were built, very rapidly, between 1283 and 1287, incorporating the *Llys* into the walls between Towers 15 and 16. This is the only section of the walls where windows are present

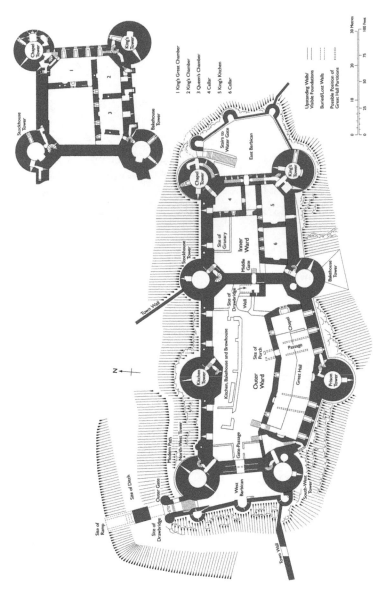

Figure 4.16 Plan of Conwy Castle. (Crown copyright Cadw (2020).)

Figure 4.17 Conwy Castle.

(though these postdate the construction of the *Llys*). Once again, the now familiar construction and planning team of Master James of St George as master mason, Richard the Engineer and master carpenter Henry of Oxford appear listed in the Exchequer accounts (Ashbee, 2006). Sir John de Bonvillars, later constable of Harlech Castle, appears to have general management of the programme of works (Ashbee, 2015). The curtain walls and towers were first constructed between 1283 and 1285. By 1285, work had also started on the town walls and on the royal apartments and other buildings in the Castle's interior (Figure 4.17).

The Castle is divided into two wards. At the west end was the main 'outer' ward which could be accessed from the town. The eastern 'inner' ward contained the royal apartments, and these could be accessed from a postern gate in the east-facing curtain walls. This private door brought the royal family and their retinue through a garden area and a path down to a fortified wharf on the riverbank. The inner ward could be accessed from the outer ward, but for extra security, these two parts of the castle interior were separated by a ditch which was crossed using a drawbridge.

The Great Hall and royal apartments remain relatively intact and unmodified (despite their current ruinous state). As such they are one of the most important examples of such buildings of this period in the United Kingdom (Ashbee, 2006). They were built between 1284 and 1286 and as well as a number of halls and receiving rooms, include a fine, circular chapel built into the most northeasterly tower, now named the Chapel Tower. The remains of beautifully constructed traceries, and therefore once glazed, windows are situated overlooking the inner ward courtyard from the eastern rooms, which, although ahead of their time in terms of design, are contemporary with the original phase of building in the 1280s (Ashbee, 2006). Edward I only stayed in these apartments once, during Christmas of 1294.

By the 16th century, when Henry VIII commissioned a survey of the Welsh castles, it seems that Conwy Castle had become dilapidated and was in dire need of repair. The stonework appears to have been in good order but the carpentry and drains were in a poor state. In 1621, similar works had to be carried out again, replacing the wooden floors of the upper stories and ceilings in general, where rotten timber presented the visitor with a 'double dainger of fallinge through the floare and havinge the beames that hange loose above to fall upon his head' (from a contemporary description of Conwy Castle quoted by Taylor, 1995).

In all phases, the castle is built and repaired from very locally sourced stone, primarily quarried from the outcrop on which it stands and within the town, with dressings in a mottled sandstone from the Creuddyn Peninsula on the north side of the Conwy Estuary. Exceptions are repairs and vaulting put in during the mid-14th century when red sandstone was imported from Chester for renovation of the interior.

The rocky, sandstone promontory, located at the confluence of the Rivers Conwy and Gyffin, limited the size and geometry of the castle that was to be built upon it. This is an outcrop of the Ordovician Conwy Castle Grit, a sandstone member enclosed within the Conwy Mudstone Formation. This outcrop is tightly folded and steeply dipping. It is also dissected by faulting, but it can be followed along its tortuous strike for around 3 km west and southwest of the castle. To the south east, the unit is bounded by the Afon Gyffin Fault. Bedding thickness is variable, ranging from a few centimetres to 30–40 cm thick beds. This has put an upper limit on the size of building stones. The lithology includes brown-weathering, blue grey sandstones, which are sometimes calcareous. "Grit' is something of a misnomer; the majority of the lithologies, though considerably coarser-grained than the surrounding mudstones, are medium-grained sandstones. Some horizons contain strings of pebbles or mudstone rip-up clasts and beds also exhibit lamination to various degrees and on various scales. Blocks are not infrequently cross-cut by thin quartz veins and occasional bedding-parallel quartz veins. Where broken along these vein structures, mineral fibres are evident.

It is largely, but not wholly, from this unit that the stone for the castle was quarried. As at Harlech, levelling the site and excavating the castle ditch to the west of the Castle would have provided a substantial amount of stone, but other quarries also existed along the outcrop of the Conwy Castle Grit. The main 'Town Quarry', of which only traces now exist, is bisected and accessed by St Agnes Road and located just to the south west of Conwy Station (Neaverson, 1947).

The Conwy Castle Gritstones break naturally into tabular-shaped blocks which are defined by the intersection of bedding planes with a regular joint pattern. Lott (2010) has noted that larger stones, extracted from the thicker bedding planes, were used to construct the round towers, whereas blocks from thinner beds were used to build the intervening walls (Figure 4.18).

The slaty, slab-like Ordovician sandstones were totally unsuitable freestones, and it was therefore not possible to produce decent quality dressings and moulded masonry features from this material. The window dressings and surrounds for arrow slits in the castle walls and towers (and to a small extent in the Town Walls) are a local but somewhat enigmatic stone, the Gloddaeth Purple Sandstone Formation. This is a coarse to very-coarse gritstone and is made distinctive by its purple-red to buff-yellow variegated colour. It exhibits well-developed trough cross-bedding with beds some 30–40 cm thick. Petrologically, this is a litharenite (Lott, 2010), dominated by quartz with feldspars and lithic fragments in a matrix of kaolinite and minor silica cement. Sedimentary structures and grain morphology indicate a fluvial environment of deposition (Figure 4.19). This stone outcrops locally on the Creuddyn Peninsula, and it seems it was almost entirely quarried out in the Medieval period, although traces of it remain in the Gloddaeth Syncline in area of Bodysgallen House at Bryn Pydew (Barnard & Collins, 2007; Warren et al., 1984). Dating of these horizons (and other

Figure 4.18 The masonry of the North West Tower of Conwy Castle. The brown stones are the Conwy Castle Grits. The dressings around the blocked arrow loop are Gloddaeth Purple Sandstone. A put-log hole is located just to the right of the arrow loop.

Figure 4.19 A block of cross-bedded and mottled Gloddaeth Purple Sandstone used as a quoin at Conwy Castle.

red-bed sequences in North Wales) has been problematic, with speculation on a potential range of Upper Carboniferous and Permian ages. Analyses by palynologists at Robertson's Research (unpublished but cited in Davies et al. (2011) have found fossil miospores which indicate that they probably belong to the Early Pennsylvanian Warwickshire Group, rather than the Mid-Pennsylvanian, Coal Measures, but place the Gloddaeth Sandstone firmly within the Upper Carboniferous.

In the 1340s, Edward the Black Prince's register records that stone was brought from Chester 'to complete the stone arches of the hall' (quoted in Ashbee, 2006). This is presumed to refer to the ceiling arches in the Great Hall in the outer ward but could also have meant or included the ceiling arches in the royal apartments of the inner ward of the castle (Figure 4.20). However, it seems that red Triassic sandstones had been imported, presumably by sea, from Chester relatively early in the construction of the castle; a fine-quality freestone was required to produce finely carved and moulded window traceries as well as the aforesaid ceiling arches. The local, coarse-grained Gloddaeth Sandstone was not suitable for such delicate mouldings. Lott (2010) has identified this stone, also used for interior fireplaces to be predominantly red Helsby Sandstone Formation (Sherwood Sandstone Group) but mottled red and yellow Triassic sandstones of the Helsby Formation are also present. These are medium-grained sandstones and would have been familiar to Edward I's builders; having lived in Chester, Richard the Engineer would certainly have been aware of the superior quality of his local rocks as building stones as they had been in use for masonry construction in Chester since the Roman period and were quarried within the town.

Unlike the other Edwardian Castles, there is considerable remaining evidence that the walls of Conwy Castle were rendered in white mortar and limewash, which would have transformed the appearance of the castle from its current dark and somewhat oppressive state into a castle befitting those seen in medieval illustrated manuscripts. Traces of plaster are seen on both interior and exterior walls of the castle and particularly obvious on the exterior of the main entrance gatehouse. Where remaining, these are composed of a lime plaster with a fine aggregate of sand and a minor amount of brick dust. White patches or 'lime lumps' suggest that a hot lime mix was used, that is, one that was used immediately after slaking and the addition of aggregate.

Figure 4.20 The area of the Great Hall at Conwy Castle with the Prison Tower on the right. The soffits and springs of the arches over the Hall are constructed from Permian Sandstone from Cheshire.

4.5.2 Conwy town walls

The town walls at Conwy are more extensive and far more complete than those at Caernarfon and with a few minor exceptions (i.e. the arch where the railway cuts through the walls and arches enlarged or added to admit road traffic) have had very little modification since they were built between 1285 and 1287. The walls are 1.3 km in length and comprise twenty-one towers, which are identified by numbers which move sequentially in an anticlockwise direction along the circuit of walls, starting at the first tower on the east end of the Quayside. A ditch was constructed along outer side of the walls. The walls and ditch enclose an area of 10 ha. Architecturally, the walls and towers are constructed in the same style as those of the Castle (which is not the case at Caernarfon) and therefore clearly represent unified architectural planning. According to Ashbee (2015), documentary sources appear to refer to both Castle and town as the *castrum* and not distinguishing them as separate entities. Architectural evidence, such as the spiralling put-log holes (for supporting scaffold) which occur in both towers on the wall and in the castle, and styles of features such as corbelling and arrow slits indicate probably the same teams of workers working on both components as an integrated whole. However, whilst there is no evidence that the walls were rendered in plaster, as was the Castle, they may have been simply painted white; construction accounts for 1286 do include payments for limewash (Ashbee, 2015).

The stone used to build the town walls is a mixture of the Conwy Castle Grit Formation sandstones used in the castle and a distinctive buff-coloured rock, a rhyolite (Figure 4.21). This stone belongs to the Conwy Rhyolite and Capel Curig Volcanic Formations rocks both similar in appearance and adjacent in outcrop which are part of the Llewelyn Volcanic Formation. The Conwy Rhyolites outcrop trending NNE-SSW from Foel Fras to Conwy, attaining a maximum thickness of ~1,000 m. On Conwy Mountain, the unit has been metasomatised, with sericite replacing feldspars. The Capel Curig Volcanic Formation outcrops over a large area from the Capel Curig area to Conwy, it represents a major ash-flow tuff, up to 400 m thick (Ball & Merriman, 1989). The Llewelyn Volcanic Group represents the earliest expression of Caradoc Volcanism in North Wales and consists of trachyandesites, rhyodacites and rhyolitic ash-flow tuffs.

At Conwy, rhyolites are mainly found in the town walls near to Bodlondeb, suggesting that the Capel Curig Volcanic Formation is the stone used here, rather than the Conwy Mountain Rhyolites. Devitrified fiammé are occasionally observed in the rhyolite blocks

Figure 4.21 The masonry of Conwy Town Walls. Dark-coloured stones are Conwy Castle Grits. The pale coloured stone is rhyolite.

of the town walls, indicating a more probable pyroclastic origin rather than that of a siliceous lava. Stone was quarried due west of the town in a band that runs up Conwy Mountain from Bodlondeb (an area located immediately west of the town walls). The rocks are ivory-coloured but with slightly greenish-tinge, fine-grained, siliceous rock, with bright orange iron-staining and sparsely por-phyritic with K-feldspar. Some surfaces show drusy quartz crys-tals, presumably formed along veins which have now been split in half. Both formations are dominated by quartz, potassic feldspar, plagioclase and muscovite.

Neaverson (1947) speculated that the rhyolite in the Town Walls of Conwy may have been reused masonry from the castle at Deganwy, which had been destroyed by Llewelyn ap Gruffudd dur-ing the reign of Henry III in 1263. This is not implausible, but no records exist either to support this theory or to dismiss it. Given the proximity of quarry sites in Bodlondeb it is just as likely it was quarried nearby rather than being collected from the ruins of Deganwy and shipped across the estuary.

4.6 BEAUMARIS CASTLE

Beaumaris Castle is situated on the coast of the Menai Straits, close to the north-eastern point of the island of Anglesey (Ynys Môn) and the Irish Sea. Unlike the other three castles, Beaumaris was built on flat, marshy land underlain by glacial deposits, and although on the coast, its overall plan was not constrained by being on a prom-ontory or river mouth. Beaumaris, meaning 'fair marshes', was a new town constructed by Edward I's planners and castle builders. This meant that the castle, the last of the four to be built, could be meticulously and symmetrically planned; Edward's 'perfect' castle. It is also the only one of the four castles to have a proper moat, in the absence of a rocky outcrop some outer form of defence was deemed to be important. Construction of the castle began in 1295, following the Welsh uprising of 1294 when Madog ap Llewelyn, a relative of the defeated Welsh Princes, led a rebellion against the English settlers and Edward I's castles. Caernarfon Castle and town were taken and set ablaze, but Harlech and Conwy could be sup-plied by sea and were able to resist attack. Edward organised a mil-itary campaign to Anglesey, the heartland of the rebels in 1295. His wrath was focused on the regional centre of Llanfaes in north-east

Anglesey on 10 April 1295 where he captured Madog and sent him to London for imprisonment. With a week, he had engaged Master James of Saint George and made a down-payment for the construction of a new castle to defend Anglesey, and one that, like Harlech and Conwy, could be supplied from the sea. The town of Llanfaes had to go. Like the old town of Conwy, it contained a port, a royal *llys* and a Franciscan friary. Edward had the town cleared and its inhabitants resettled in Newborough on the south coast of Anglesey. The new town of Beaumaris was to be located a mile south of the site of Llanfaes.

Beaumaris Castle was built between 1295 and 1306, and a huge amount of money was invested in the first 6 months, a time during which Edward was living on site. However, the funds soon dried up. In 1296, Edward, seemingly a man who could not endure peace, began a war with Scotland, an endeavour that earned him the epithet *Malleus Scotorum* (the Hammer of the Scots), and this campaign would occupy him until his death in 1307. Money was taken away from the Beaumaris project despite the pleas of Master James and Walter of Winchester, who in 1296 wrote in desperation to the exchequer for money to pay the workers' wages; 'and sirs, you should know ... we owed more than £500 for the workmen and the garrison. We are finding it very hard to retain these men who have been and are still in such great want because they have nothing to live on'. (quoted in Ashbee, 2017b). Despite the pleas, the money was not forthcoming, and as a result, the Beaumaris Castle was never completed.

As at the other castles, the building accounts survive for Beaumaris Castle and provide a huge amount of detail concerning the people who built the castle (Taylor, 1987; Knoop & Jones, 1932; Ashbee, 2017b). In their letter sent at the end of 1296, Master James and Walter of Winchester state that at the time they were employing 1,000 carpenters, plasterers, blacksmiths and diggers as well as 200 quarrymen plus boat crews to transport stone, 400 masons, 2,000 general labourers and a garrison of 130 ft soldiers, cavalry and bowmen. These figures illustrate the large size of the workforce and say something about the division of labour in the construction of a late 13th century castle.

The plan of Beaumaris Castle is symmetrical, along a north-south axis, and it was surrounded entirely by a moat, of which around half still exists (Figure 4.22). There are two sets of concentric walls. The inner ward was constructed during the initial phase

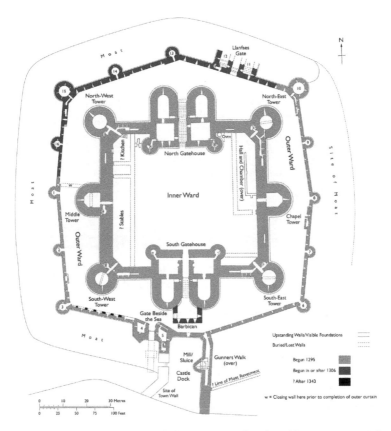

Figure 4.22 Ground plan of Beaumaris Castle. (Crown copyright Cadw (2020).)

of building 1295–1296 and is enclosed by a more or less square set of walls, consisting of double-turreted North and South Gatehouses and curtain walls with six additional towers. The outer ward is enclosed by a hexagonal wall with sixteen towers including a double-turreted outer gate the 'Gate beside the Sea' which was accessed by a drawbridge crossing the moat. This outer wall was not fully completed until the second phase of building which began in 1306 which also included a gate on the NNW wall, the Llanfaes Gate, which was never completed. A barbican was added to the South Gatehouse of the Inner ward at this time (Figure 4.23). The relatively soft glacial deposits that the Castle in built upon have

Figure 4.23 Beaumaris Castle. The fields in the distance were once occupied by the town and priory of Llanfaes.

been responsible for subsidence over the years, and a number of vertical fractures have opened up which are visible in the outer curtain walls of the Inner ward, particularly around the North Gatehouse (Figure 4.24).

The building stone of Beaumaris Castle has been recorded in a number of works, including a pioneering survey by Edward Greenly (1932) and then by Neaverson (1947). More recently, Lott (2010) has published the results of a survey conducted by the British Geological Survey. Despite proximity to good-quality stone at Penmon, which had been worked in Roman times, the castle is predominantly constructed from gritstone, laminated limestones and the local high-grade metamorphic basement. The local limestone would have been used for making lime mortar and plasters.

According to the building accounts drawn up by Walter of Winchester, the cost of stone, which includes quarrying, breaking and dressing and its transport to the site is a significant cost in the expenditure (Taylor, 1987). This suggests that the majority of the stone was indeed quarried fresh rather than being *spolia* from abandoned buildings in the town of Llanfaes (though it is likely only the friary church would have been stone built at the time). Transport of 'stone' from 'the quarries' is frequently mentioned in the building accounts, though little if no information on the location of the quarries or the type of stone is provided. Some stone was bought in

Figure 4.24 The East Wall of Beaumaris Castle, near the Llanfaes Gate. Fractures caused by subsidence in the underlying marshy ground at Beaumaris.

'divers places' perhaps suggesting that not all was fresh quarried. Stone was bought in from subcontractors, and a Robert of Preston and Alan of Kirkeby are mentioned as the leaders of gangs of boatmen and stone-breakers. Robert's men brought and split flagstones. These were brought by sea and so possibly refer to the flaggy limestones rather than the local schists. Frustratingly only three types of stone (*petrarum*) are named in the accounts, and they have proven to be largely undeciphered; *petrarum velosarum, petrarum de Maylloun* and *petrarum de Portwynge. Portwynge* is most probably Porthaethwy (Menai Bridge). Accounts refer to 6,000 freestones hewn from the quarry at Portwynge, and this would fit the squared blocks of Anglesey Grit observed in the Castle walls. Maylloun, clearly also a place name, remains an enigma. It is possible that it may refer to Moelfre or even Amlwch? The potential source of limestones and gritstones. However, it bears little resemblance to any modern Welsh local place name, though the closest word in modern Welsh is *llwn*, meaning 'dust'. *Petrarum velosarum* probably refers simply to 'rough stone',[10] which could arguably be applied to any of the stones used in the construction of the castle.

10 *Villosus* means 'rough' or 'shaggy' in Medieval Latin (Ashdowne et al., 2018).

Lower Carboniferous limestones are the main building stone used in the castle, both in the walls of the Inner and Outer Wards. However, it is somewhat surprisingly not the good-quality 'marble' facies stone quarried at Penmon or the grey Cefn Mawr Formation ashlars used at Caernarfon Castle (although these stones are used in subordinate amounts). This is a laminated and in places almost shaley limestone, a lagoonal facies rich in pyrite and a dark blue-grey in colour when fresh, but weathering to a dull, dark rusty-brown. This lithology belongs to the Leete Limestone Formation of the Clwyd Limestone Group, possibly quarried at Tandinas, a quarry just west of Penmon Point. However, Greenly believed that the particular facies used in Beaumaris Castle was quarried between Benllech and Moelfre on the north side of Red Wharf Bay (Greenly, 1932). This may be the *Maylloun* stone referred to in the building accounts (Taylor, 1987). A 'quarry five leagues distant' is also mentioned in the building accounts, and this could indeed be the one referred to, though it can be located with no certainty (Neaverson, 1947).

As at Caernarfon, Anglesey Gritstone is widely used at Beaumaris Castle. Squared ashlar blocks of this buff-coloured, coarse-grained, Basement Beds sandstone are used in a chequer-board pattern, in contrast with the darker-coloured, Tandinas Limestone on several towers, though use of this pattern is far from consistent, implying a fluctuating supply of well-cut stone of both lithologies (Figure 4.25). This decorative use of stone is particularly apparent on the upper courses of the South-West and Middle Towers of the Inner Ward and therefore appears to have been something of an afterthought. As at Caernarfon, decorative stonework indicates that there was never an intention to render the castle walls with plaster. The gritstone is also used for lintels and for dressings around windows and arrow loops. This stone is also used for the carved heads of gargoyles used on the castle latrines and other drainage features coming from the walls. Gargoyles of this type are not seen in any of the other Edwardian Castles and are a unique feature of Beaumaris. The gritstone varies from a medium to coarse-grained sandstone of the Malltraeth-type to gritty, very poorly-sorted, breccia-conglomerates containing pebbles of vein quartz and prominent granules of red jasper. Although the building accounts for Beaumaris suggest that quarries near Menai Bridge (*petrarum de Portwynge*) were worked, Greenly firmly believed the majority of this stone came from the Penmon and Benllech areas of Anglesey and was quarried alongside the limestones. It is of course very possible and extremely

Figure 4.25 The outer wall and South West Tower of Beaumaris Castle. A chequerboard pattern can be seen on the South West Tower, made of contrasting pale Basement Beds gritstones and darker Leete Limestone. This carefully laid stonework is in contrast to that used on the outer walls.

likely that several sources were utilised, as this lithology outcrops widely in coastal Arfon and on Anglesey.

The coastal plains of Anglesey, bordering the Menai Straits, are underlain by the Neoproterozoic Penmynydd Terrane which consists of predominantly epidote-glaucophane schists, as well as lenses of garnet-glaucophane schists. The chlorite-epidote schists and related rocks from the Penmynydd Terrane, and particularly the Pen-y-Parc Formation of this unit, are well- and finely foliated and deformed by small-scale folds. Stones quarried from this unit are not uncommonly used in the walls of Beaumaris Castle and are an extremely distinctive and unusual choice of building stone. Its attraction lay only in the fact that this stone outcropped very locally and broke with relative ease, albeit into somewhat ragged slabs. Such stones have never been considered prime contenders for ashlar masonry. However, their occurrence in Beaumaris Castle brings unexpected interest for the metamorphic petrologist as well as being, generally, an uncommon building stone (Figure 4.26).

A few other stones occur as small components of the overall structure. Infrequent examples of high-quality limestone ashlar do occur in the castle, and it is odd that this facies had been

Figure 4.26 A block of greenschist in the walls of Beaumaris Castle.

overlooked in favour of the shaley and overall lower-quality Tandinas Stone. These Penmon Quarry facies are pale grey and often show the characteristic bioturbation or well-preserved corals and brachiopods and are a mixture of Cefn Mawr and Loggerheads Limestone. Limestone blocks are well-cut ashlars and also occur as slab-like blocks used for the lintels of windows and doors in the walls. Groups of limestone blocks of different sizes are used, again a feature which suggests the reuse of ready-cut stone (Figure 4.27). It is not impossible that these stones represent *spolia* from a few well-built structures in the former town of Llanfaes or were even collected from an unknown Roman building, rather than their being specially quarried for construction of the Castle.

Rounded boulders and cobbles, gathered from either the underlying glacial tills or from the beach, are also used sporadically within the structure, as well as infrequent use of white quartz-arenites and other sandstones that are probably been derived from the Creigiau facies of the Lower Carboniferous Basement Beds and even rarer examples of local Ordovician grey sandstones. However, these stones are uncommon in the general fabric of the Castle, and their use does not indicate systematic quarrying of this rubble. The beach origin of several of these stones is obvious from the presence of borings made by the piddock bivalve (*Pholas* sp.; Figure 4.28). Cobbles occur predominantly in the wall cores along

Figure 4.27 The west-facing inner wall of Beaumaris Castle showing patches of Cefn Mawr and Loggerheads Limestone, which contrast with darker-coloured Leete Limestone. Groups of blocks of uniform sizes are used, including a string of small setts which can be seen on the centre left of the photograph. These purposely cut blocks indicate the re-use of cut stone obtained from other buildings.

Figure 4.28 A Pholas bored cobble, collected from the local beach and used in the fabric of Beaumaris Castle.

with rubble of the rock types described above but also occasionally are found in the facings of walls.

4.7 CONCLUDING REMARKS

Often seen as a unified building programme, organised by the same team of architects and engineers and ultimately designed and overseen by King Edward I, the four Edwardian Castles that comprise the UNESCO WHS are each individual and unique in terms of their architecture, design and building materials. This is perhaps more surprising given that three of them, Caernarfon, Conwy and Harlech Castles, were being built simultaneously, albeit in areas with different geological resources. It was certainly not unusual during this time period for stone buildings to be constructed of very locally sourced stone. However, the use of stone in the Edwardian Castles is sometimes enigmatic, with potentially good sources of local stone (such as the Penmon Limestones of the Loggerheads and Cefn Mawr Formations) being overlooked in favour of seemingly intractable and unwieldy stone such as greenschists and conglomerates, as is the case at Beaumaris Castle. This and previous surveys of the stones used in the construction of the four castles have still been unable to answer some of the questions raised concerning the use of *spolia*, stone recycled from older buildings that were demolished. Of particular interest here are the Roman fort at Segontium and other buildings mentioned (or not, as the case may be) as sources of masonry in the contemporary building accounts such as Deganwy Castle and Maenan Abbey. Good limestone ashlar, from the Cefn Mawr Formation, is used at Caernarfon (and to a much lesser degree at Beaumaris). Could this be because ready-quarried and cut stone was taken from the Roman fort of Segontium or perhaps from other, yet unknown and unexcavated (or completely dismantled) Roman buildings hidden beneath the urban environment of Caernarfon?

Further work is required to begin to answer these questions, but this would require an extensive study that is well beyond the scope of this study. A thorough archaeological survey of the quarries, especially the limestone and sandstone quarries of the Lower Carboniferous of Anglesey, could shed much light on their periods of use. Temporal variations in quarrying tools and techniques and comparison of these with extant buildings, not just the Castles but also churches and other regional stone buildings of the later Medieval period (and into the modern period), may begin to answer questions regarding choices made when selecting stone. Similarly, a thorough campaign of measurement and study of building block sizes at Caernarfon Castle, Segontium and other Roman forts and structures may throw some light on the degree of spoliation of Roman buildings in the early stages of the building Caernarfon Castle.

Study of the Castles has also highlighted the geologically poorly understood lithologies that were important in the 13th and 14th centuries as castle building stones and have subsequently been quarried out and are therefore absent from geological maps and the scientific literature. These are Egryn Stone, Gloddaeth Sandstone and the distinctive, red Lower Carboniferous sandstones quarried from Moel-y-Don and to a certain extent, the Ordovician Allt Llwyd Grits used at Caernarfon Castle. The Castles are clearly historically unique, and assumptions relating to this status may be readily applied to their building materials. However, to put these and other local and imported building materials in a more regional context, it is necessary to look at the stones used in vernacular structures throughout the region. This is the purpose of the next chapter.

Chapter 5

Building the towns of North West Wales
Churches, civic centres and manor houses

5.1 HISTORIC BUILDING IN NORTH WEST WALES IN THE 14TH TO 17TH CENTURIES

The towns of Caernarfon, Conwy, Harlech, Beaumaris and also Criccieth were new towns established by Edward I between 1283 and 1295 and initially populated by migrants from England (Lilley, 2010).[1] However, apart from street plans, which largely remain the same as they were in the early 14th century, there is little evidence besides the castles and their town walls and churches, of buildings from this time. However, in the century following the construction of the towns, Wales went into a period of decline. Like that of much of Europe, the Welsh population decimated by the plague in the 14th century. The Plague entered Wales in 1349 and within a year, a quarter of the population had died. Records for NW Wales are scarce, but Davies (1993) speculates that the scattered populations of the uplands of Snowdonia were less badly affected than other areas. However, outbreaks of disease affecting livestock had also hit the region badly in the early 14th century which would have had an enormous impact on the livelihood and survival of farmers and farm labourers, who would have been the majority of the population at this time. The plague recurred throughout the century,

---

1 Elsewhere in Wales, Edward I also established the towns of Aberystwyth, Rhuddlan and Flint in 1277, Denbigh, Ruthin, Holt and Fflint in 1282–1283 and Caerwys and Overton in the 1290s. He also established Newborough in the early 14th century as a town for the residents displaced from Llanfaes for the establishment of the new town of Beaumaris.

DOI: 10.1201/9781003002444-5

further reducing an already struggling population. There is little, if any, evidence of stone building of this period in the region.

Owain Glyndwr (Owain ap Gruffydd) was descended from pre-conquest Welsh Royalty. He was born in the mid-14th century and trained as a lawyer in London, but he had also gained considerable military experience and a good grasp of political acumen. By 1399, a combination of events and circumstances, not least the instability wreaked by the plague in the second half of the 13th century and the resultant isolation and poverty of the Welsh population, led Owain to believe that the time was right for the Welsh to revolt against English dominance. In England, this was also a time of political turmoil. The English King Richard II was usurped by Henry Bolingbroke, the son of John of Gaunt, Duke of Lancaster, who was declared King Henry IV. This disruption of the royal line of succession and the tumult associated with this change of Royal houses gave Owain the opportunity to rebel against English Rule. Owain Glyndwr declared himself as Prince of Wales in 1400, with a promise to deliver the Welsh from oppression. Many Welsh migrants to England returned home to join the cause, and Glyndwr also received support and endorsement from the Cistercian Monasteries, and indeed Englishmen who found dissatisfaction with the emergence of Henry IV's rule. By 1405, Charles VI of France was also giving Glyndwr support (Davies, 1993). Battles and skirmishes took place over the next few years; however, by 1408, Henry IV was gaining the upper hand. Glyndwr had occupied Harlech Castle in 1408, but the following year the Castle was retaken by the English and Glyndwr fled into the mountains. It is likely that he died in c. 1415, though, surprisingly, little is known about the final years of his life.

English histories tend to describe Wales as a broken and lawless place for much of the ensuing 15th century, but literary and architectural evidence in fact suggests a general improvement in the economy. During this time, the construction of large and well-funded churches flourished in NW Wales, including St Mary's and St Nicholas's Church in Beaumaris and major alterations to St Deiniol's Cathedral in Bangor as well as the spectacular St Beuno's Church in Clynnog-Fawr. The remarkable Dovecote at Penmon was also built around 1600. In England, following years of war with the Lancastrians, the House of the Yorkist Kings was on its last gasp with the Reign of Richard III. Without a clear successor, the crown was seized by Henry Tudor after the defeat of Richard at the Battle of Bosworth in 1485. Henry Tudor, King Henry VII, was in part of Welsh descent, but his claim to the throne was through his English bloodline. Nevertheless, many of the Welsh population identified with Henry Tudor (perhaps more than Henry Tudor identified with the Welsh) and during his reign Wales and England became unified and prosperity increased in the region. During the 15th and 16th century, large merchants' houses were built in the towns of Conwy and Caernarfon as well as country manors and a number of churches in the English Perpendicular architectural style, the most impressive of which is St Beuno's at Clynnog-Fawr. Henry Tudor's successor, the tyrannical Henry VIII brought about religious reformation in England and Wales, establishing Protestantism and demolishing the power of the Cistercian order in Gwynedd as well as their monastic buildings, including the foremost religious complex at Maenan in the Conwy Valley.

Following the construction of Conwy Castle, Maenan Abbey and monastery had been established in the Conwy Valley, 3 km north of Llanrwst, on the east bank of the river in 1284. The old Cistercian abbey had been located at the mouth of the River Conwy, but this was in the way of Edward I's building plans for the new castle and the monastic community was relocated upriver. Unfortunately, there are no records concerning the extent of the building works and excavated remains are fragmentary. It has been assumed that building works reused stone from the Roman Fort at Canovium (at Caerhun), though it is not clear how many stone structures were present there in the Roman period. The Abbey was dissolved and dismantled in 1538, and the stone was reportedly used to build nearby Gwydir Castle and also transported to Caernarfon Castle to make repairs (Piers, 1916). Detailed dissolution records remain from 1539 to 1540 (Owen, 1917); however, there is no obvious use of stone from the Conwy Valley region in use at Caernarfon, either in the Castle or in town walls. Stone was almost certainly reused locally for buildings including Gwydir Castle and, possibly, a late 18th century manor built on the site. The current Maenan Abbey Hotel replaced this building in 1848, and it has been a hotel since the 1960s. Excavations in the hotel grounds in 1963 uncovered only twenty-two pieces of worked stone, all unidentified sandstones, and a great number of robbing trenches; the latter suggesting that stone had indeed been reused elsewhere. Butler and Evans (1980) state that 'no stone merits illustration', but they describe it as 'fine-grained white, grey or pink, with a few coarser red sandstones'. Butler and Evans (1980)'s assumption was that this stone was brought from Cheshire, but it is more likely to be the Gloddaeth Purple Sandstone, quarried at Deganwy, as was used in Conwy Castle and in Medieval buildings in Conwy and also occurs at Gwydir Castle (Kenney, 2012; Haslam et al., 2009). A quarry near Maenan in the Denbigh Grits, presumably opened for the construction of the Victorian Maenan Abbey manor, is shown on the 1890 Ordnance Survey map. It is not shown on earlier maps from 1788. Although there is poor evidence of reuse of stone, the story of Maenan Abbey, through contemporary reports of its demolition and archaeological investigation show that in a landscape rich in accessible stone, cut and dressed stone was valued and reused.

Churches and other places of worship are often the oldest and best maintained buildings in communities worldwide. This is the case in North Wales, where churches were often stone built and have been well maintained, and for the most part, sensitively restored or renovated. The Victorian architect, Henry Kennedy, who was employed by the Diocese of Bangor, did much to preserve the ancient churches of Arfon and Anglesey. Churches are also buildings of considerable local prestige and patronage and have been constructed from good-quality stone. As such they are a good place to assess the quality of the locally available buildings stones over time. Even when church buildings are restored and original stone is reused, new stone is often introduced to replace decayed stonework. Some of the most ancient churches in the region, built before Edward I's conquest of Wales, have been already discussed in Chapter 3. The important churches in the landscape associated with the Edwardian Castles are discussed below with relevance to their associated towns. The somewhat isolated, but spectacular church of St Beuno, in terms of both its stone and its architecture, is discussed here within the context of Renaissance North Wales.

Clynnog-Fawr is a small village with a very big church located mid-way between Caernarfon and Pwllheli, and this is probably the most impressive building dating from the late 15th century in the region (Figure 5.1). Reputedly the burial place of St Beuno, one of the founding fathers of the Celtic Church, Clynnog-Fawr was the official starting point of the Medieval pilgrim road to Bardsey Island. Beuno is said to have died in c. 650 CE at Clynnog-Fawr (Llanfeuno) where he had founded a monastic house on land gifted to him by Cadwallon, King of Gwynedd (Sims-Williams, 2004). Like the majority of the Welsh Saints, Beuno was not a martyr. He

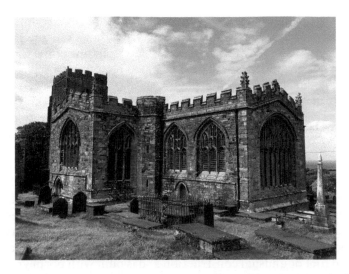

Figure 5.1 St Beuno's Church at Clynnog Fawr. The church is built of local porphyry microgranite with window and door dressing of various sandstones, including Egryn Stone.

reputedly lived for almost 100 years, was abbot of his monastery and performed a number of miracles, raising at least seven people from the dead during this time. One of these was his niece, the nun Gwenfrewi (later St Winifred[2]), whom he managed to resurrect despite her death by decapitation. The site is associated with several early Christian inscribed stones which have been discussed in Chapter 3. The current St Beuno's Church was built on the scale of a cathedral in the late 15th century, in the English perpendicular architectural style. The main church is connected to the small Capel-y-Bedd (the shrine built over St Beuno's grave; *bedd* means 'grave' in Welsh) by a roofed passageway. Construction is believed to have begun around 1480, and the building was completed in the early 16th century, the tower and the Capel-y-Bedd are probably the latest structures to have been built. The main building stone is the local porphyritic microgranite which would have been quarried on Gyrn Ddu, a kilometre to the south-west of the church.

---

2 St Winifred is also associated with a church and sacred well at the eponymous Holywell in Clwyd.

A number of disused quarries occur of the west-facing slopes of Gyrn Ddu which were in operation up until the 19th century and a series of inclines are still in place. Gyrn Ddu and its satellite intrusion at Bwlch Mawr are the easternmost plutons of the Caradoc Nefyn Cluster Intrusions. The Gyrn Ddu pluton is a porphyry microgranodiorite, with 5 mm plagioclase laths set in a glassy grey matrix mottled with darker clots of chlorite. It weathers to an attractive, pale-pink colour. Orthogonal joint sets enable thin, but potentially large (1–2 m) slabs to be extracted from the quarries, which are used in the randomly coursed, rubble masonry construction. The foundation is notably of built from large, rough-hewn, and perhaps glacially derived boulders of the porphyry that would have been collected as field stones. The passage connecting the church tower to the chapel is of interesting construction, the architecture being driven by the properties of the building stone. The passage is roofed by a series of large porphyry slabs above which is a corbelled roof of unusual construction and is built with porphyry rubble and glacial boulders. The connecting passage from the Tower to the chapel was probably made in the 17th century. This building is certainly the largest in this region to have been constructed of this porphyritic microgranite.

Dressings around doorways and window traceries at St Beuno's are also made from sandstone. A mixture of sandstones with different textures, colours and weathering characteristics have been used. These range in colour from buff, pink, pale blue-grey to yellow with prominent Liesegang bands. Dark red and liver-coloured sandstones represent more recent phases of restoration and repair. The original dressings are made from either various medium grained sandstones or gritstones from the Lower Carboniferous Basement Beds and Egryn Stone from the Harlech Dome. The square-arched, west-facing door in the tower and the window above this are constructed from a distinctly pink and mottled red and ivory sandstone. This includes dressings with head stops in the form of bishops wearing mitres. This is Basement Beds sandstone from Moel-y-Don on the Menai Straits shore on Anglesey, the 'peaches and cream' facies of Davies (2003). A few blocks show intense weathering, picking out cross-bedding and forming tafonic cavities. This weathering is unusual for Basement Beds sandstones but would have been exacerbated by the west-facing aspect of the door and the proximity to the sea and salt-laden winds. The same stone is used on the west-facing doorway of the Capel-y-Bedd.

The main entrance porch at the north-west corner of the nave is also dressed in sandstone, but the large amount of lichen has made identification of the stone used difficult. A clearer picture is seen in the doorway in the north transept. The upper part of the arch is red, liver-coloured sandstone which is probably a repair, but the lower arch is constructed of a yellow sandstone with prominent Liesegang banding. The door jambs are of a blue grey sandstone. Both the blue and yellow colouration, plus the presence of iron spots and Liesegang banding in the yellow variety are typical of the range of colours observed in Egryn Stone which was quarried in a limited region to the south of Harlech and used in Harlech Castle and known to have been exported as far away as central Ceredigion (Palmer, 2007). A door with similar materials used for dressing is also situated on the south transept. This is currently the most northerly known example of the use of Egryn Stone.

In addition, courses of red sandstone moulding are seen in the lower parts of the church walls. This is much weathered and encrusted by lichens but is a red medium-grained sandstone from the Basement Beds outcrops at Moel-y-Don or Pwllfanogl quarry areas on the Menai Straits of Anglesey. Dark red and liver-coloured cross-bedded, Triassic Cheshire sandstone has been used more recently for replacement stone in areas requiring repair.

St Beuno's Well is located 500 m west of the church, and the basin is now surrounded by a small, unroofed well-house. It is built from the same pink-weathering porphyry microgranite as the church.

On Anglesey, Plas Coch (Red House) is a striking 16th century mansion, which is currently privately owned and part of a holiday resort. It is located in Llanedwan, just inland from the Moel-y-Don quarry site on the Anglesey shores of the Menai Straits. The original house was built by a lawyer, Dafydd Llwyd in 1569, and substantially enlarged by his son, Hugh Hughes, in the 1580s. Hughes became attorney general for Wales. The house was further extended in the 19th century (Haslam et al., 2009) and again in 2010 to include a pool and leisure complex. The original mansion has a plan in the shape of a letter E, which was very much the fashion during the reign of Queen Elizabeth I, with three wings projecting at the front façade, each surmounted with crow-stepped gables. As the name suggests, it is built from a bright red gritstone, the local Basement Beds Grits which were quarried on site (Campbell et al., 2014). A range of coarse gritstones and conglomerates varying from medium-grained red and mottled red and cream-coloured sand-stones are present in the building. Window dressings and a plaque over the front door are carved from the freestone facies of buff-coloured Basement Beds quarried at Malltraeth. Nineteenth century embellishments, including garden walls and gate posts, are in Triassic sandstones brought in from Cheshire. A modern extension, built by David Insall Associates, is clad in Lower Permian, red Penrith Sandstone (Lazonby Stone) from north western England. Plas Coch Lodge, though built in local Basement Beds red gritstones, and also with crow-step gables, was also built in the 19th century. Both the lodge and main house are roofed with purple-coloured, Cambrian Heather Slate, probably from Penrhyn.

There are relatively few 17th century buildings remaining in the region. However, a network of thriving towns were well established, and this is evident from the cartographer John Speed's maps of the North Wales (and indeed the rest of the British Isles). John Speed (c. 1552–1629) was one of the principal cartographers of the early 17th century, and he published his map of Wales in 1610 and then county sheets in the following year. The county-scale maps include as insets plans of the principal towns in each county. From these we are able to see street plans and the principal buildings in place at Caernarfon, Bangor, Beaumaris and Harlech at this time. Outside the towns, the main buildings from the 17th and 18th centuries are vernacular; farmhouses and associated barns and stone-built houses in villages and towns. Locally derived stone from small quarry pits or field stone is primarily used in these contexts, which given the variation in geological materials therefore gives a certain local character even within the relatively small area under study. Basement Beds gritstones and sandstones continued to be widely used. In the areas around Caernarfon and the eastern Llyn Peninsula, large glaciated boulders some up to 1.5 m derived from the thick blanket of boulder clays have been roughly dressed and are used as the main building stone. This is the main construction stone used in the villages of Llanrug, Bontnewydd, Groeslon, Pen-y-Groes, Dolbenmaen and Criccieth and for the estate buildings at Glynllifon. The dominant lithologies present are sandstone, some of volcaniclastic origin and green-blue coloured volcaniclastic sedimentary rocks often developing a white weathering crust. These boulders are also ubiquitous in the construction of field walls. An exception to this norm is the town of Beaumaris (see below) which was in this period the administrative centre and principal town of

Anglesey and what is seen today is essentially a Georgian Town, with elegant buildings constructed of limestone ashlar.

In the 18th century, North Wales became increasingly wealthy, with the establishment of country seats in the ownership of the local gentry, many of whom grew rich in the agricultural and quarrying industries. Some of these were local families, with members rising to be members of parliament or achieving other senior military and administrative positions and further rose through the ranks through good works and fortuitous marriages. By the later 18th century, huge wealth was beginning to be generated from the slate industry, and this hugely strengthened the positions of these families in the landscape of North Wales. The second half of the 18th century was also a time of political instability in Europe and, for Britain, war with the former colonies in North America. By the end of the century, war with France and Napoleon Bonaparte was on the horizon.

5.2   NEW CASTLES OF THE 18TH AND 19TH CENTURIES

The only substantial structure to be built with a genuine defensive purpose since the construction of Beaumaris Castle is Fort Belan, which guards the western entrance to the Menai Straits. This small fort was built by the first Lord Newborough in the late 18th century. Newborough's estate, Glynllifon, now a country house hotel located nearby in Llandwrog, is one of the four main country seats in the region along with Plas Newydd, the home of the Marquis of Anglesey (now managed by the National Trust). However, grand houses were also built by and for the slate barons who found themselves in possession of increasing wealth at the beginning of the 19th century. Indeed the Lords of Newborough owned quarries and leased land for quarrying in Dyffryn Nantlle which helped to fund Plas Glynllifon. The Vaynol Estate in Bangor belonged to the Assheton-Smith family who owned the Dinorwig quarries in Llanberis. These buildings, and their estates, for the most part, fit the mould of the standard British stately home. Penrhyn Castle, belonging to the Pennant family who owned the enormous Penrhyn Quarry at Bethesda, is a 19th century, over-the-top, architectural novelty, and although arguably a castle in the true sense of the word, its defensive qualities were intended only for dramatic effect.

5.2.1 Fort Belan

Fort Belan was built with the intention of defending the narrow entrance to the Menai Straits and the town of Caernarfon from the perceived threat of invasion during the American War of Independence in 1775. It was constructed by local gentleman Thomas Wynn of the Glynllifon Estate (see below), later to become first Lord Newborough who needed a more effective defensive site to house his Caernarfonshire militia, who had previously garrisoned the tiny Fort Williamsburg on the Estate. Having said that, Fort Belan is not a large structure. It consists of a rectangular fortress with barracks enclosed by stone ramparts housing canons. Adjacent is a small, hairpin-shaped dock surrounded by one-storey warehouses. Fort Belan is currently in private ownership.

The fort is located on the northern tip of a peninsula separated from Caernarfon by the shallow Foryd Bay. This spit of land has formed as the result of long-shore drift and is composed of recent wind-blown sand overlying sandy and muddy tidal flat deposits, which ultimately overlie Ordovician basement consisting of Nant Ffrancon Group clastic sediments.

The Fort is solidly built of Basement Beds gritstone rubble, with brick dressings. However, many of the buildings are also rendered in concrete. The dock also has copings of gritstone ashlar. The warehouses surrounding the dock are built from gritstone rubble along with boulders of glacial field stones, grey and dark green in colour, derived from the volcanic hinterland of Snowdonia. All the buildings in the fort and dock area have purple Cambrian slate roofs, presumably using slate from the quarries in Dyffryn Nantlle which were owned by the Wynn Family.

5.2.2 The Vaynol Estate

The Vaynol Estate, containing the New and Old Halls and extensive parkland, is located almost entirely on Carboniferous Limestones of the Clwyd Group and underlying Basement Beds gritstones. These strata are bounded to the west by the NNE-SSW trending Dinorwic Fault, placing these sedimentary rocks in a graben with respect to the basement Cambrian and Ordovician strata of mainland and Precambrian rocks of Anglesey. The Carboniferous strata include the so-called Bridges Formation of Basement Beds grits, which here are variably coloured from yellow through green, red and lavender and also containing exotic clasts of Mona Complex greenschists. These are overlain by the locally named Treborth Limestone Formation. The Treborth Limestone is now correlated with the Loggerheads Limestone and Cefn Mawr Formations of the greater Clwyd Group. Four main beds of gritstone are interbedded within this formation. Characteristically, the limestones from the Vaynol Estate are strongly stained pink due to the once overlying red beds of the Upper Carboniferous Westphalian Warwickshire Formation (locally known as the Plas Brereton Formation). These strata outcrop along the coast south-west of Vaynol, between Port Dinorwic and Caernarfon.

The Vaynol Estate belonged to the Assheton-Smith family, owners of the Llanberis slate quarries, who also developed their own slate port, Port Dinorwic (Y Felinheli) which is located on the Menai Straits directly south west of the estate. Originally the land belonged to the Bishop of Bangor, but in the mid-1500s, it became the seat of the High Sherriff of Caernarfonshire William Williams. It remained in the same family for over two centuries. Much of the estate remains in private ownership and is not accessible to the general public; however, the building stones of the notable structures are unusually well detailed in Pevsner's Architectural Guide (Haslam et al., 2009). Notable buildings on the estate are the New and Old Vaynol Halls and a chapel. The most obvious and well-known feature of the Vaynol Estate is its imposing perimeter wall built in c. 1860, 5 km of Loggerheads Limestone quarried from Moelfre and capped with purple Dinorwic slate (Cambrian slate) coping stones. The name Vaynol comes from the Welsh *Y Faenol* meaning a manor. The Old Hall dates to the 16th century and is similar in design to Plas Goch on Anglesey, an E-shaped Elizabethan manor house constructed of Basement Beds gritstone roughly coursed, rubble masonry. These stones range in colour from red to grey and are quarried from the Menai Straits Formation, most likely from quarries on both sides of the Straits, the redder stones probably coming from the Moel-y-Don and Pwllfanogl quarries. Dark red, liver-coloured sandstones are used as for a number of the large quoins on the corners of the building. The window and door dressings are of a fine-grained sandstone freestone, possibly the facies of Basement Beds quarried at Malltraeth on Anglesey. The roof is of purple Cambrian slate. A small chapel of the same period is adjacent to the Old Hall and constructed of the same materials.

The New Hall was built in 1793 by architect James Defferd of Bangor for the Assheton-Smiths. Roofed with Dinorwic slate, the walls and frontage are rendered with white plaster so other building materials remain unknown, and little is known about the construction history, the record of accounts having been lost in a fire at another of the family's houses (Haslam et al., 2009). Other buildings of note on the Estate are a mausoleum built in the Gothic architectural style and designed by the local ecclesiastical architect, Henry Kennedy. Constructed in 1878, Pevsner informs us that this is built from 'limestone rubble, lined with Bath stone with Mansfield stone columns'. (Haslam et al., 2009). The limestone is surely the local pink Treborth limestone. Bath Stone, a middle Jurassic oolitic limestone from the Great Oolite Group was, and still is, a well-known and routinely used English building stone. Mansfield Stone comes from the Permian Cadeby Formation (Magnesian Limestone). A rare clastic rock in a series of Zechstein dolomites is composed of detrital quartz in a dolomitic cement. Quarrying at Mansfield in Nottinghamshire began in 1337, under a licence from Edward III for use in the construction of Southwell Minster. Mansfield is the source of Red and White Mansfield Stone and the compact dolostone known as Mansfield Woodhouse Limestone. The White Mansfield sandstone was quarried at the south end of the town of Mansfield in the Maun Valley. It was favoured because it carved easily when fresh and hardened with age (Lott, 2001).

5.2.3 Plas Newydd

Plas Newydd is located directly opposite the Vaynol Estate on the Anglesey shore of the Menai Straits. It too sits on Cefn Mawr Formation Limestones of the Clwyd Group. The site of a Medieval Hall was extended in several building phases over 500 years. In the mid-18th century, Plas Newydd was in the hands of the Member of Parliament for Anglesey, Sir Nicholas Bayly (d. 1782). Bayly married Caroline Paget, the daughter of a Baronet, and their son Henry Paget inherited the baronetcy and also became the first Earl of Uxbridge. Henry's son, Henry William Paget, was second in command to the Duke of Wellington at Waterloo, and he was granted the title of Marquess of Anglesey in 1815, following the British victory at the battle. The house has been the seat of the

Paget family ever since but is now in the hands of the National Trust. Henry Paget was responsible for the transformation of the house to the building we see today (Figure 5.2). It was built in two main phases in the later 18th century, the first under the direction of local architect, John Cooper of Beaumaris between 1783 and 1786, the second phase from 1793 to 1799 under the direction of English society architect and master of the gothic style, James Wyatt, who was assisted by Joseph Potter. The main building stone is pale weathering, dark grey Cefn Mawr Limestone, reputedly from Moelfre (Haslam et al., 2009). This stone has a fauna of sparse, shelly fossils, including brachiopods, gastropods and the rare goniatites (see Figure 2.12d).

The interior of the house is most famous for the *trompe l'oeil* mural, which depicts a fantasy seascape and harbour, in the dining room by early 20th century artist Rex Whistler. There is much of interest to see in the architecture and history of the house and its, on occasion, colourful occupants. The main item of note to the connoisseur of decorative stone is the Adam-style fireplace in the SW Bedroom, attributed to Benjamin Bromfield by Haslam et al. (2009). It is in white Carrara Marble, with panels of yellow Siena Marble on the uprights and a central panel of Cotham Marble in the crosspiece. Cotham Marble is from the Rhaetic (Upper Triassic)

Figure 5.2 Plas Newydd is built from Cefn Mawr Limestones.

Cotham Member of the Lilstock Formation, was popular as a British decorative stone and is often seen in *pietre dure* work and on specimen marble tabletops. A 'landscape marble', it is an algal biohermal travertine; Cotham Marble locally forms domal stromatolites up to 20 cm in height. They are cut by desiccation cracks which are responsible for the formation of the structures which resemble trees and hedges in the polished cross sections.

Henry William Paget was also commemorated by the Marquess of Anglesey's Column, located in Llanfair PG. Constructed of Loggerheads Limestone from Moelfre, the column is sited on Cerrig-y-Borth, an outcrop of glaucophane-garnet schist (and a protected Site of Special Scientific Interest; SSSI, Campbell et al., 2014). A single Doric column is surmounted by a statue of the Marquess in bronze, the whole structure is 27 m in height. The architect was Thomas Harrison, and it was completed in 1860 (Haslam et al., 2009).

5.2.4   The Glynllifon Estate

Another extensive Estate is located in the parish of Llandwrog, just south of Caernarfon. Glynllifon belonged to the Wynn Family who owned Cambrian slate quarries in Dyffryn Nantlle and extensive land in the region but also had stakes in slate quarries in the Conwy Valley and Blaenau Ffestiniog. Sir Thomas Wynn (1736–1807) was the first Lord Newborough, and he was succeeded by his son Thomas John Wynn (1802–1852). The large park house, Plas Glynllifon, was designed on a monumental scale by Edward Haycock and completed in 1846 and standing on the foundations of an earlier manor built in the 1750s. Much of the frontage was originally whitewashed, but it is now mainly rendered in concrete. Beneath the render, the main building stone is Loggerheads Limestone from Penmon and Moelfre, visible in the enormous, pedimented portico at the front of the house, supported by six ionic columns. The roof is purple Cambrian slate, surely sourced from Dyffryn Nantlle.

Geologically, the majority of the Glynllifon Estate is sited on Pleistocene glacial and fluvio-glacial deposits. Cambrian Arfon Group sandstones and slates of the Fachwen Formation outcrop in the northern part of the Estate, and this unit underlies the Ice Age deposits. A large part of the glacial deposits are push-moraines, which are well exposed in section on the coast beneath the Iron Age hill fort of Dinas Dinlle (Harris et al., 1997). The park house itself is situated on fluvio-glacial sheet-flow deposits. Both these units were sources of a substantial number of large, glacial, erratic boulders, largely derived from the local hinterland of Snowdonia, and these are typically green to grey blocks of Ordovician volcanic and volcaniclastic rocks. Large glacial boulders have been used for decorative effect in the landscaping of the park and for constructing estate buildings.

To the north of the house is a square-plan block known as the Estate Yard. The buildings enclose a courtyard recently paved with bricks and gravel. This complex of buildings was constructed from 1836. The buildings use the large (some very large, with long-axes dimension of 1–1.5 m) glacial boulders. These have been roughly dressed and shaped into ashlars and laid in roughly coursed masonry. Close examination of these stones shows that they are predominantly green and green-grey coloured, volcaniclastic lithologies, some mineralised with pyrite. Also present are blocks of a brown sandstone facies. Cambrian, purple Nantlle slate is used for lintels on both doors and windows and for roofing. The broad, carriage arches are dressed with red brick. The buildings here have been recently restored and converted into a visitor centre and shops and craft workshops.

5.2.5 Penrhyn Castle

The overly impressive sprawl of Penrhyn Castle dominates views of the coastal plain north of Bangor, in the village of Llandegai. The underlying bedrock is Ordovician Nant Ffrancon Group clastic sediments, sandstones and shales. The park is cut by a north west trending, though far from straight, dolerite dyke, and it is on the low ridge that this feature has created that the castle is built. This enormous building was largely constructed between 1820 and 1837 by the architect Thomas Hopper for George Hay Dawkins Pennant (1764–1840), who had fortuitously inherited the estate of his father's cousin, Richard Pennant (1737–1808). Dawkins-Pennant, an owner of 764 slaves across four plantations in Jamaica, had also inherited the slate quarries at Penrhyn. He was also a member of the Committee of the West India Association which did not support emancipation. Nevertheless, when emancipation happened in 1833,

Dawkins-Pennant was at the time compensated almost £15,000 for his lost slaves,[3] an equivalence of over £1 million pounds in terms of modern expenditure, money that must have come in very useful in the completion of building works on his castle. The money from the plantations had also produced considerable capital which was used to expand the family's slate concerns into what is still the largest slate quarry in North Wales. The workmen at his quarries at Penrhyn worked in atrocious conditions in an environment where the threat of accidents was ever present. The Castle was orientated so that the Penrhyn Quarry was well framed as viewed from Pennant's study in the keep.

Often regarded as an over-the-top expression of 19th century wealth and capitalism, Pevsner praises Penrhyn Castle's 'Neo-Romanesque' architectural style, particularly in the quality, its uniformity of style and the attention to detail in the various aspects of the building (Haslam et al., 2009; Figure 5.3). Penrhyn Castle is built in the style of a Norman keep, the like of which had never been seen before in this region, which has no shortage of castles. It is all-rounded, chevron-decorated arches and arcading and fanciful

Figure 5.3 Penrhyn Castle is constructed primarily of Loggerheads Limestone.

3 Legacies of British Slave-Ownership, UCL History: Accessed 6 August 2020, https://www.ucl.ac.uk/lbs/.

carving. The architect Thomas Hopper (1776–1856) was also at around the same time building Gosford Castle in Armagh, Northern Ireland in the same style. Hopper had specialised in building country houses and had also worked at Windsor Castle and Carlton House for the Prince Regent (later George IV).

The building stones of Penrhyn Castle are relatively well recorded, and a display case in the study houses some examples of building stones, though their labels and provenance should not be too heavily relied on. The main building stones are Clwyd Group Limestones quarried at Penmon and Moelfre on Anglesey (Haslam et al., 2009; Marsden, 2009) and specifically from the Loggerheads and Cefn Mawr Limestone Formations. The materials display notes that Flagstaff Limestone was used. Flagstaff Quarry is located just to the north west of Penmon Priory on the cliffs overlooking the northern end of the Menai Straits and exposes Loggerheads Limestone.

The interior fittings and fabric of Penrhyn Castle contain some interesting and noteworthy use of local stone, as well as imported materials. Upper Carboniferous York Stone from the Pennines Lower Coal Measures Group of Lancashire and West Yorkshire is used and polished for flagstones in the Great Hall. These stones have taken on a lustrous appearance here which is in significant contrast to their more familiar use as exterior, street paving stones. The impressive structure of the Grand Staircase is a Neo-Romanesque flight of imagination on behalf of Thomas Hopper. A 12th century Norman castle would have had spiral staircases, but artistic licence was required to adorn Penrhyn Castle with a more dramatic form of access to upper storeys. This square-plan staircase makes a series of turns amid a framework of carved sandstone and decorative plasterwork. The walls of the staircase shaft are clad in an ivory-coloured, sparsely fossiliferous, uniformly textured oolitic limestone. This is Painswick Stone, blocks of which were recorded as being delivered to the building site in the 1820s (Marsden, 2009). Painswick Stone is an important English freestone which is quarried from the Middle Jurassic (Bajocian) Cleeve Cloud Member of the Birdlip Limestone Formation in Gloucestershire. One of the Cotswold stones, it is a relatively rare and unexpected use of English Jurassic oolite in North Wales. Hopper had previously worked at Windsor Castle, and contemporary records state that Painswick Stone was used in works there, showing that it was becoming used outside its local area by the 1820s (Brewer, 1825).

The balustrades, rails and newel posts of the staircase are flamboyantly carved from a grey to pale yellow brown, fine-grained sandstone. The origin of this sandstone is unknown, but it is almost certainly one of the grey sandstones from the Pennine Lower Coal Measures Group of Lancashire (Marsden, 2009). The most likely contender is the iron-poor massive sandstone of the lower sections of the Old Lawrence Rock Member which outcrops around the towns of Wigan and Parbold in East Lancashire and known today as Appley Bridge Blue Stone. The spectacular carving afforded by Hopper's masons demonstrates that this stone has good attributes as a freestone; the crown-shaped newel posts are decorated with characters performing various knightly and chivalric pursuits. The upper part of the Old Lawrence Rock is both more flaggy and more iron-rich and has ripple laminations which concentrate organic matter. This or a very similar PLCM sandstone has been used on the stair rails and elsewhere in the castles as dressing stone.

A number of substantial stone chimneypieces and other examples of decorative stonework are placed in certain rooms in the castle. Many of these are in locally derived decorative Anglesey 'marbles'. The 'marble' facies of the Penmon Limestone (Loggerheads Limestone) is capable of taking a high polish and is a bioturbated, mottled, blue-grey to brown limestone known as Penmon Marble. This stone was widely used throughout the British Isles as a decorative stone, especially for ecclesiastical fittings and for decorative stonework. There are four chimneypieces made from Penmon Marble in the Library of Penrhyn Castle and another in the Breakfast Room. Penmon Marble is also used for decorative console tables in the Great Hall and elsewhere throughout the castle.

The Dining Room has a chimneypiece of black marble. Black Marbles (mostly bituminous limestones) are notoriously difficult to identify when lacking in fossils and are available from several locations in Ireland, Belgium and South Wales. Thin beds of bituminous limestones outcrop within the Clwyd Limestone Group at Penmon and Dinorben and a black marble quarry is recorded at Dinorben (see Daniel & Ayton, 1814; Horák, 2011). A century later, Watson (1916) records a sample from the Dinorben Quarry in the University of Cambridge's Sedgwick Museum which is 'deep blue-black in colour and may safely be classed among the Black Marbles of the British Isles, although it cannot be compared with the deep black marbles of Ireland or the continent'. Nevertheless, it is very likely that this is the stone used in the dining room of Penrhyn Castle (Jana Horák, *Personal Communication*).

The most spectacular chimneypieces in Penrhyn Castle are those constructed of so-called Mona Marble which are situated in the Drawing Room, the Slate Bedroom and the Study in the Keep. Mona Marble is not a true marble in the geological sense, this stone is a purple-red serpentinite, cross-cut with 'leak-green', red and white veins with some areas showing brecciated textures. Outcrops of this serpentinite occur in the deformed New Harbour Group of the Mona Complex on Anglesey. The main outcrop occurs in a series of pods along strike from Mynachdy, SW to Llanfechell and is predominantly a red or purple ophicalcite with only local green serpentinites. The serpentinites are associated with metamorphosed gabbros and other ophicalcites. According to Watson (1916), this stone has been occasionally used in ecclesiastical trappings, but its use had declined at the time of his writing. It has been used in Bristol, Peterborough, Truro and Worcester Cathedrals. The stone was mainly used for chimneypieces. The exotic sounding 'Verde de Mona' (the green variety) was exploited and promoted by George Bullock in the early 19th century, a time when the Napoleonic Wars prevented European marbles and serpentinites being traded into the British Isles. The red serpentinite fire-surrounds at Penrhyn Castle were supplied by a Thomas Crisp from quarries near Llanfechell (Horák, 2005, 2011; Llwyd, 1833). Chimneypieces in the servants' quarters, as well as counter tops in the kitchens and dairy are all made from Penrhyn Slate. The famous slate bed in the Slate Bedroom, probably carved by a local craftsmen or gravestone carver, is also in Penrhyn Slate.

Mona and Penmon Marbles are used as polished table tops in various locations around the castle, along with other imported decorative marbles, including Connemara Marble from the Republic of Ireland, Tholonet Breccia from Provence, France and a much-battered slab of *Breccia policroma di San Benone*, a stone of unknown origin and probably once a grand tour souvenir, perhaps acquired from an archaeological site, collected by a member of the family.

5.3 EIGHTEENTH AND NINETEENTH CENTURY TOWNS

5.3.1 The coming of the railways

Local Welsh slate, and particularly that from the Cambrian Slate Belt (Llanberis Slate Formation), is universally used as a roofing material across the region. The slate industry began to develop on an industrial scale in the later 18th century, bringing a new and prosperous industry to the NW Wales in the early 19th century which required manpower and the development of infrastructure, slate works and ports, to support the quarrying activity and the global demand for roofing slate.

Slate brought considerable wealth to the quarry owners, and some of this trickled down to the regions, with the need to employ quarrymen and engineers, overseers and foremen as well as dock workers and ships' crews. Housing and infrastructure were required to support the expanding population and the towns began to grow. The slate, quarried inland, up in the mountains had to be transported down to the coast for shipment abroad. Early horse- and gravity-drawn railways were installed in the quarries, connecting the inland, upland areas of slate production with the ports on the coast.

NW Wales was also gaining importance to travellers as it was on the road to Holyhead and the sea crossing to Ireland. This brought travellers to the region who initially needed new and improved roads and then eventually a rail network as well as facilities for lodging. In the 19th century, the region also grew in importance as a tourist destination, with seaside resorts established at Llandudno and Harlech and at other coastal towns and in the mountain resorts at Betws-y-Coed and Beddgelert. The Castles of Edward I also became important localities on the tourism map.

The railways connected the region to the rest of the United Kingdom and connected the ferry link to the Republic of Ireland from Holyhead on Anglesey. By 1848, a line from Chester to Bangor had been built and another from Llanfair PG[4] to Holyhead on Anglesey. These lines were connected by the completion of the Britannia Bridge crossing of the Menai Straits in 1850. The line was further extended to the slate ports of Y Felinheli and Caernarfon,

---

4 Llanfair PG Station, or to give it its full name, Llanfairpwllgwyngyllgogerychwyrndrobwllllantysiliogogogoch, is famous for being the station with the longest placename in the United Kingdom. Llanfair PG is a shortened version of the village's most used names, Llanfairpwll or Llanfairpwllgwyngyll. The medieval settlement was called Pwllgwyngyll. The railway sign, still affixed to the old station house, is considered to be more of a tourist gimmick rather than a regularly used toponym. It translates as 'St Mary's *llan* (church) at the pool of the white hazel trees near the fast whirlpool and the *llan* (church) of St Tysilio's at the red cave'.

and this opened in 1852. In 1870, from the Caernarfon Station, located at Doc Fictoria, a branch line was extended through the town (the Carnarvon Town Line), along the course of the ancient River Cadnant (now Crown Street) and through a tunnel to the station on the Slate Quay. This allowed connections with lines bringing slate from Llanberis and also from Rhyd Ddu from where the railway continued to Porthmadog and connected with the Ffestiniog Railway. Porthmadog was connected by a railway line to Harlech in 1860. This railway network provided a fully integrated regional transport system for slate, quarrymen and passengers. The Caernarfon-Porthmadog line has recently reopened as the Welsh Highland Railway which runs to Porthmadog and the Ffestiniog Railway and operates primarily as a tourist line. The main rail network now has its railheads at Bangor and Holyhead.

5.3.2 Building bridges

The main obstacles to both road and later, rail travel on the Chester-Holyhead route were the crossings of the Menai Straits and the estuary of the River Conwy, and the construction of bridges in the early 19th century represents some of the greatest engineering works of this period anywhere. Geologically, the narrow Menai Straits are formed by the influx of the sea into a graben structure which separates the Island of Anglesey from the mainland, mainly governed by the NNE-SSW trending Menai Straits fault system and primarily defined by the Dinorwic Fault (Gibbons, 1987). There were always crossing points by ferry, and in the 13th century, during Edward I's campaigns in the region, pontoons were built in at least two places to facilitate the crossing of the fast-flowing and often treacherous waters. The first, permanent road crossing, the Menai Suspension Bridge was constructed between 1812 and 1826 under the direction of the pioneering civil engineer, Thomas Telford. The bridge has a total span of 427 m, but the span of 178 m between the two towers was at the time the longest in the world. At 100 m above water level, the bridge had to be high enough to accommodate the tall masts of the slate ships.

The Menai Suspension Bridge is constructed of Penmon Limestone (Loggerheads and Cefn Mawr Limestone). The stone was contracted by Telford via a Mr. James Wilson at a cost of £165,000.

The iron suspension chains were made in Shropshire by William Haszledine. The same team of engineer, stone contractor and iron worker were also responsible for the Conwy Suspension Bridge (1822–1826) which crosses the mouth of the River Conwy connecting the Castle rock with Deganwy, a total span of 101 m. Again, Penmon Limestone was the main building stone. The new road was fully completed in 1832, enabling a much easier journey to the ferry port of Holyhead on Anglesey and connections to Ireland.

Two decades later, the main Chester to Holyhead road was superseded by the railway. This time, Robert Stephenson was the engineer responsible for securing the crossings, with the construction of the tubular rail bridges over the River Conwy and the Menai Straits. Conwy Railway Bridge was constructed between 1846 and 1848 and has a total span of 123 m, with towers and buttresses, again, of Penmon limestone Figure 5.4. The Britannia Tubular Bridge, crossing the Menai Straits, was a far more ambitious structure. A huge amount of detail concerning the construction of the bridges from an engineering point of view, and especially the testing of the iron tubes which would take the steam engines (a separate one for each direction), was published by Stephenson's chief engineers, Sir William Fairbairn (1849) and Edwin Clark (1850). These volumes are highly recommended to any reader with the intention of building their own tubular railway bridge but unfortunately apply scant evidence to the selection of stone used for the bridges, though it is stated that the stone was tested for its load-bearing properties. It is oft quoted (though the precise source is obscure) that rubble core of the pylons of the Britannia Bridge was sandstone from the quarries at Runcorn and the casing was limestone from the Penmon quarries on Anglesey (see Haslam et al., 2009). Nevertheless, these are indeed entirely reasonable choices of stone. Although not visible, and therefore not verifiable, Runcorn Sandstone, a Triassic red sandstone from the Cheshire Basin already been used in engineering contexts, would have been acquired from Weston Quarry (Watson, 1911). The exterior cladding is certainly consistent with Loggerheads and Cefn Mawr Limestone from Penmon. As a final decorative flourish, sculptor John Thomas (1813–1862) was commissioned to carve four lions from Penmon Limestone and two of these were placed to guard each end of the Bridge. Sadly, the Britannia Bridge lions are now difficult to access, sited below the level of the very busy, modern road. Nevertheless, they are still there lurking in the undergrowth.

Figure 5.4 The three bridges crossing the Conwy Estuary as seen from Conwy Castle. Far left is the modern road bridge built in 1974. In the middle is Thomas Telford's Suspension Bridge and right is Robert Stephenson's Tubular Bridge. This is the railway bridge with a tube to carry trains in each direction. Note the causeway constructed from the Deganwy shore and projecting halfway across the Estuary. This is constructed on sand bars which are exposed at low tide. The bridges span the main, deep water river channel.

5.3.3 The railways bring new stones

An oft overlooked advantage of rail networks was their capacity to bring non-local stone to a region and especially that used in relatively small amounts for often decorative purposes, such as the plinths of statues, and decorative interior fittings for churches and civic buildings. Indeed across the United Kingdom the popularity of granite as a building and decorative stone grew alongside the development of railway lines put in from London to Aberdeen (1850), from London to Penzance in Cornwall (1852) and London to Dumfries and ultimately Glasgow (1859). This connected the United Kingdom's main granite producing regions with the rest of the United Kingdom, and the use of polished granite took off across the country as a decorative stone. The technique of polishing granite was invented by the stonecutters Newall's of Dumfries in the 1840s, and polished examples of the local Criffell-Dalbeattie

Granite were shown at the Great Exhibition of 1851. Within a few years, Aberdeen had taken over as not only a centre of granite quarrying, but also of granite cutting and polishing. Stone was imported by sea from the Baltic Region and even South Africa to Aberdeen where it was cut and polished before being exported further. Because of the growth of this industry, a range of Scottish and more exotic stones became readily available and increasingly in demand. These granites were used for shop fronts, church fittings and also decorated tomb memorials and grave markers. Cornish Granite, though used widely as a building stone in London, was not widely imported further north and west. It is generally only seen in North Wales as gravestones and war memorials.

A number of firms were established in the second half of the 19th century, which exploited the current trend in both architecture and the decorative arts for highly decorated surfaces and embellishments and particularly in the use of coloured stone in both interior design and for sculpture. The most famous of these businesses was Farmer & Brindley Ltd. William Farmer and William Brindley were both stone carvers and later quarry entrepreneurs. They became partners in the 1860s and set up a business in both importing decorative 'marbles' from Europe, North Africa and the United States and boosting a demand for decorative, polychromy stonework which fitted perfectly with the Victorian vogue in resurrecting the 'Gothic' and Italianate styles of architecture. Firms such as Farmer & Brindley, and also competitors J. Whitehead & Son Ltd. and Fenning & Co. Ltd., transformed the use of decorative stone in British architecture. This trend for polished and richly coloured stonework fuelled a mission to discover and rediscover British and Irish stones which could match the beauty of those derived from Italy, Greece and further afield. The interiors of civic buildings, churches, hotels, public houses, shops and homes were transformed, at least in the mind's eye of the time, to an equivalent splendour of that of Ancient Rome by the installation of decorative marble cladding and other fittings such as pillars, urns, finials and balusters.

The world 'marble' to the stone merchant was applied in the same way as it was in the 1st century CE Rome (*marmor*) to mean a soft stone which was capable of taking a good polish. This includes 'true' marbles in the geological sense but also includes limestones and indeed, serpentinites, within this category and included the 'Mona Marble' serpentinites described above.

5.3.4   Slate steps up to the challenge

At around the same time and in direct competition to the decorative marble industry, businesses specialising in the production of enamelled slate began to appear in North Wales and also in England. Slate had been traditionally used throughout Wales and across the wider British Isles for the construction of fire surrounds, amongst its many other uses. What could this monochrome stone do to compete with the current trend for fancy marble fireplaces? The technique of slate enamelling, in order to transform its appearance to that of brightly coloured and intricately patterned marbles, was invented in the mid-19th century, but very much pioneered by George Eugene Magnus (1801–1873), who leased a number of quarries in North Wales and elsewhere. Magnus used Ordovician Aberllefenni Slate from Merionethshire for his enamelled work as not all slates could withstand the necessary heating and cooling processes required to set the pigments. He too exhibited his work at the 1851 Great Exhibition. However, many other slate enamelling manufactories were set up in the slate ports and slate works, such as Inigo Jones's works at Groeslon near Caernarfon (Jones, 2006). Magnus placed an advert in The Builder in 1847 which states that enamelled slate … [is] 'now taking the place of marble in the mansions of the Nobility generally. It is also extensively used in Government Offices, the principal Railway Stations and other public buildings'. In reality it was used in the lesser railway stations and in the homes of the aspiring middle classes for whom the price of a genuine marble chimneypiece was a luxury too far. Although slate enamelling work continued well into the 20th century, in the 1860s, real marble had become much more readily available and therefore affordable and marble-effect enamelled slate was in decline.

Slate could still be a decorative material, even when in its natural state. The multi-coloured Cambrian Slates worked from Bethesda to Nantlle were particularly popular for producing patterned roofs, this style of decorative architecture was popular in eastern Europe, particularly so in Germany, but this was also a local phenomenon. Slate was also a material of huge importance in the education of the youth of the British Isles, and supplying every school child in the British Isles and beyond, with a writing slate, was a lucrative business. From the 18th through to the mid-20th century, slates were used by school children for writing practice and arithmetic. Factories such as the Britannia Slate Works at Pwllfanogl on the

Menai Straits shore of Anglesey existed solely for the manufacture and framing of writing slates and slate pencils. They were exported all over the world via rail and sea from Port Dinorwic.

5.4   NORTH WELSH WEATHER AND NATURAL STONE

North Wales records some of the highest rainfalls in the British Isles, and it is constantly buffeted by the prevailing Atlantic south-westerly storms moving up the Irish Sea. Surprisingly, slate does not have ubiquitous use for building cladding, though it certainly does occur in this context, but usually found only on west- and south-west facing, exterior walls to protect against the prevailing winds. In a few cases, houses are entirely clad with slates, particularly the Cambrian slates which come in a range of colours, to provide a decorative effect. Similarly multi-coloured and intricately shaped slates were also used to create patterns in roofing (Figure 5.5).

Harling is a mixture of flint pebbles mixed with cement which are sprayed onto walls and this material is ubiquitous in the towns and seaside resorts of north-west Wales and has been used from the 19th century to the present day. By the 1930s, harling had become

Figure 5.5 Bridge Cottage in Porthmadog is completely clad in dark grey, Blaenau Ffestiniog slate, including hexagonal slates used for decorative purpose.

the main type of weather-proof coating used on both domestic and civic buildings, obscuring old stone and brick structures and much more. Harling-coated surfaces can be painted, and indeed these white-washed or colourfully painted buildings have added to the character of north Welsh seaside towns. In addition to harling, the more sensitive use of lime wash (whitewash) has been to protect a number of vulnerable buildings, including old churches, especially those in seaside positions, lighthouses and historic buildings such as the Jacobean Plas Mawr in Conwy.

5.5 MODERN BUILDING: A RETURN TO WELSH STONE

The slate industry went in a huge decline following strikes in the early 19th century and then the impact on the workforce of the First and Second World Wars. However, a renewed interest in the use of natural and local stone has occurred during the later 20th and 21st centuries, and once again, the slate industry is viable, but at a much smaller scale than before. In the Cambrian slate belt, quarries just to the south of Penrhyn are currently being worked, as is Cwt-y-Bugail and Llechwedd at Blaenau Ffestiniog. All concerns are now operated by Welsh Slate Ltd., part of the Breedon Group consortium of quarries. The days of the slate barons are well and truly over.

Besides slate, quarrying continues in many other parts of North Wales, and the acquired stone is used locally and further afield. The Clwyd Limestone still continues to be quarried at a small scale from Moelfre and Benllech on Anglesey and also in the Abergele and Halkyn regions of Clwyd in north east Wales. They are predominantly used for lime production but a small amount of dimension stone and the 'marble' facies (Penmon Marble, Halkyn Marble) are still worked for new builds and conservation work. Granodioritic igneous rocks are still worked at Penmaenmawr and at Trefor and Nonhoron on the Llyn Peninsula. Dolerite is quarried as aggregate from Minffordd near Porthmadog. Quaternary deposits are worked for sand and gravel in the eastern Llyn Peninsula at Bryncir and Pen-y-Groes as well as near Criccieth (Cameron et al., 2020). Boulder clays have also sporadically worked at Caernarfon for brick production, though at the time of writing there is no activity in the quarry.

There has been a clear move for new developments to employ Welsh Stone in many new buildings, though this is not exclusively the case, for example, German Jurassic limestones have been used for the new arts centre at Bangor University. The South Welsh stone Pennant Sandstone is quarried from the Rhondda Member (Rhondda Beds) of the Pennant Sandstone Formation, part of the Upper Carboniferous Warwickshire Group. Quarried at Gwrhyd near Neath, it is becoming widely used both as paving and for dimension stone. The visitor centre at Conwy Castle is built from Pennant Stone, and it is used as a particularly attractive paving material around Harlech Castle. The stone is often referred to Gwrhyd Pennant or Blue Pennant Stone.

5.6   THE TOWNS OF NORTH WALES: NATURAL STONE AND LOCAL CHARACTER

A review of the buildings stones used in the main urban centres of NW Wales, including the larger towns associated with Edward I's Castles, is provided below. They provide a wider view of the regional changes in building stone preference throughout the area under study and provide a wider context for the use of stone beyond the construction of the Castles and the town walls. The towns are discussed in alphabetical order.

5.6.1   Bangor

Bangor is a university town and a cathedral city and one of the principal civic centres of NW Wales. For these reasons there are a wider variety of building stones used here than the smaller castle towns and regional centres. Bangor Cathedral, more properly, The Cathedral Church of St Deiniol, is the oldest building remaining in the town which was established in the 6th century when Deiniol became the first Bishop of Gwynedd (Figure 5.6). Bangor may therefore be the oldest Bishop's seat in England and Wales. The town is sited slightly inland in the Adda Valley, itself formed by a north-west trending fault which dissects the sub-crop of the Minffordd and Bangor Formations comprising Cambrian volcanic and volcanic-lastic rocks. These strata form the ridges that enclose the town; the River Adda is now culverted beneath the town (RCAHMW, 1960). The low-lying town centre stands on Ordovician Nant Ffrancon

Figure 5.6 Bangor Cathedral. (The Cathedral Church of St Deiniol.)

Group sedimentary rocks. The cathedral stands in this valley, not visible from the sea and therefore hidden from Viking raiders. Apart from the 13th century Cathedral, little remains of the Medieval settlement. An exception is a chimney stack (made from Basement Beds gritstone and possibly dating to the mid-15th century) which has been re-erected on a 19th century building on the site of the former Plas Alcock town house. Although a few buildings have foundations dating from the 17th century (RCAHMW, 1960), the majority of the town's structures date from the early 19th century as Bangor first became developed both as a slate port and as an important staging post in the road route from Holyhead to Dublin with the construction of the nearby Menai Suspension Bridge. The university (University College of North Wales, UCNW) was established in 1884.

Although built on the site of a monastery founded by St Deiniol in the 6th century, much of what is seen of Bangor Cathedral actually dates to the mid-19th century, due to a major phase of restoration by Sir George Gilbert Scott which began in 1868 and was completed by his son, John Oldrid Scott in 1884 (Haslam et al., 2009). Scott was a great believer in restoring Medieval buildings back to the original plan and structure, or at least as closely as his understanding of the original design and fabric went. Gilbert Scott certainly paid attention to detail in Bangor and organised excavations around the church to discover its original 13th century footprint. He also endeavoured to remove all 16th to early 19th century restorations. The 13th century church was probably cross-shaped and similar in plan to the current nave and crossing, but with an apse at the east end. The earliest structure preserved in the fabric is a wall of the early 13th century building which is now part of the south wall. The 13th century church was largely rebuilt by Scott reusing original stonework and window traceries (including those dating from the 15th and 16th centuries). The west tower was added in 1532, and the tower over the crossing was added as late as 1971 by Alban Caroë.

The main building stone of the Cathedral is Basement Beds grits with subordinate limestones from the Clwyd Group, laid as coursed rubble masonry. A wide variety of Basement Beds gritstones are used, probably quarried from numerous sources. These range from fine to medium-grained pink, red and mottled sandstones coming from the Anglesey Menai Straits shore at Moel-y-Don and Pwll-fanogl. Coarse and very coarse-grained grits and conglomerates, rich in vein quartz and clasts of red jasper, are more typical of the Lligwy formation exposed at the base of the limestones at Penmon and Benllech. Colours of these stones range from red through yellow to green. A few examples of a greenschists, given a knotted appearance due to large quartz augen, are also present. These stones are typical of the Basement Beds exposures of the Bridges Formation, exposed on the mainland shore, close to Bangor between the Britannia and Menai Bridges and would have been found as weathered out blocks on the shore and within the quarried gritstones from this locality. Blocks of carbonate rocks are scattered through the construction, again as coursed rubble. These are a variety of limestones, but mainly grey-weathering Cefn Mawr porcellanous limestones and shaley, dark grey, rust-weathered muddy limestones of the Leete Limestone Formation. A few pieces of slate and local bedrock are also dotted throughout the rubble masonry fabric.

The dressings around the main door in the west front and that opening on to the south aisle are Tudor and have dressings of a soft, very weathered red and white mottled, cross-bedded sandstone. These are probably Triassic sandstones from the Helsby Sandstones of Cheshire. The 16th century West Tower is also largely constructed from ashlar masonry of red, liver-coloured Cheshire sandstone, probably the Keuper Helsby Sandstone. Much of this stonework is deeply weathered. Caroë's crossing tower, dating to the earliest 1970s, is constructed of black weathering, yellow Cefn Sandstone with parapet and merlions in greenish-tinged Gwespyr Sandstone (which in contrast to Cefn Stone, does not weather black; Haycock, 2018). The roof is Penrhyn Slate. The merlions around the roofline relate to this phase of building too. Now completely blackened, the stone is not identifiable from ground level but again is almost certainly Cefn Stone.

Local Cambrian slates are used as roofing materials on the crossing tower[5] and paving. A course of purple slate, laid as flat slabs, is used as a damp-proofing bed on the footings of the buttresses. Purple spotted slate is also used for paving outside the west door.

In terms of building materials, the general character of the town, with the notable exception of the Cathedral and the university buildings, is one predominantly built from brick with the great majority of buildings coated in harling. A few early 19th century buildings have escaped this weather-proofing treatment and are shown to be built of rough limestone blocks in coursed rubble masonry. A variety of Anglesey-sourced, Clwyd Group limestones are observed in these buildings, many with good examples of fossil corals and productids. The stone was easily shipped across the Menai Straits from Penmon. Of more interest is a terrace of early 19th century cottages on Glanrafon square which are constructed in course rubble masonry of black-weathering, volcaniclastic sandstones from the Bangor Formation. On Holyhead Road, in Upper Bangor, Our Lady and St. James's Church (by Henry Kennedy and built in 1866) is also built of the local stone, a striking, red-coloured volcanic breccia from the Minffordd Formation.

The original university campus was planned and designed by Henry T. Hare. The main building a double-quadrangled range of buildings with a prominent tower was built on the ridge of Minffordd and Bangor Formation volcanic rocks on the west side of town, the area known as Upper Bangor (Figure 5.7). The main university building, completed in 1911, dominates the town and is built of yellowish Cefn Stone, brought in from the Ruabon area of NE Wales. At the insistence of Hare, the roof is of Gilfach Slate, shipped all the way from Pembrokeshire in South Wales. Quarried near Llangolmen in the Presceli Hills, these are Arenig-age

---

5 The cathedral is roofed with lead.

Figure 5.7 The University of Bangor buildings in Upper Bangor. They are built of Cefn Stone from Wrexham and roofed with Gilfach Slate from Pembrokeshire.

volcaniclastic rocks with a moderate cleavage. Although poorer quality than the local slates, their 'rustic' appearance appealed to Hare. However, repairs used in the 1990s used sage-coloured slate from Nantlle (Gwyn, 2015). The North Wales Heroes Memorial Arch (1923), also in Cefn Stone, is located on Deiniol Road below the main university building. The 1950 Capel Emaus, by architect B. Price Davies, also uses Cefn Stone.

The range of University buildings built in the earliest 20th century includes the Memorial Building (School of Natural Sciences), brick-built and dressed with Portland Stone Whitbed. This and the former post office building (1909) further down Deiniol Road are rare uses of this Upper Jurassic, oolitic limestone from the Isle of Portland in Dorset in the region.

Later 19th and early 20th century domestic architecture in Bangor is in the majority brick-built and coated with harling. Of note are the early 20th century terraces on Garth Road (Gambier, Garfield and Gordon Terraces) which are built of yellow Caernarfon Brick with gateposts of red, Triassic Cheshire sandstone. Cheshire sandstone is also used for dressings at the British School (1858) and at the old Friars School Building and at the Capel Tabernacl. In the latter two instances, red and white, mottled Cheshire Sandstone is used. Friars School is otherwise built of Penmon Limestone, and the Tabernacl is built of grey, porphyritic Trefor Granodiorite.

Buildings constructed over the last 50 years have, on the whole, seen a return to the use of local building materials. New shopping centres and supermarkets are clad in local slate, whereas the Deiniol Shopping Centre and the Police station are built from Moelfre Limestone, with a rich fossil fauna of corals and brachiopods. Below the main university building is the Pontio Arts Centre, constructed in 2016 and faced with German Jura Marble, a Jurassic sub-lithographic limestone from the Treuchtlingen Formation and rich in fossil ammonites, belemnites and sponges.

Bangor has always been important as a port. The old port area of Garth features a number of rubble walls constructed from discarded ship's ballast. Here a jumble of one-off stones, mostly of unknown and varied provenance, are used to construct garden and other boundary walls and include a variety of granitic and other igneous rocks, green amphibolite-facies metamorphic rocks, Irish Lower Carboniferous limestone and even chunks of glassy, black metal-processing slag (Figure 5.8). Irish Newry Granodiorite and Wicklow Granite, both a major source of ballast (which is also seen in the repair works to Caernarfon Castle), are also used widely for walling throughout the town. On Holyhead Road, a length of 19th

Figure 5.8 Ballast used to construct walls in the vicinity of Garth in Bangor. An assortment of stones, predominantly granites, of local or exotic provenance are used in these walls. Note also the harling-coated terraces behind with roofs of Cambrian slate.

century rubble wall is constructed of Newry Granodiorite with scattered blocks of black, vesicular basalt, the latter probably also ballast derived from the Rhineland of Germany (Germany and the Baltic region were a major importers of Welsh slate). It is punctuated by an arched gateway in Penmon Limestone (Loggerheads Limestone). Port Penrhyn itself, located across the bay, just north of Penrhyn Castle was constructed in 1820 and enlarged in 1850 by the Pennant family as their port for the export of Penrhyn Slate from Bethesda. It is built of dressed Loggerheads Limestone from Penmon.

5.6.2 Beaumaris

'Beaumaris' (also Biwmares, as transliterated into in Welsh) is a French name meaning 'beautiful marsh' (*beau marais*). As the name suggests, the town is sited on salt marshes at the northern end of the Menai Straits. As the location of Edward I's final and architecturally finest castle, Beaumaris was established as a borough under royal charter in 1296 and subsequently remained the principal town of Anglesey until Llangefni was established as the county town in the 1880s.[6] This process, in true Edwardian style, destroyed the pre-existing township of Llanfaes and relocated its Welsh residents to Newborough (literally the new borough) in southern Anglesey. A medieval street plan was quickly set out, though early buildings with the exception of the Castle and possibly the church would have been timber-built and thatched. Unlike at Caernarfon and Conwy, this was not initially a walled town; however, walls were built in the early 15th century though these no longer survive. They are clearly in evidence on John Speed's map of the town published in the early 17th century, running from the Castle along the coast (the area now known as the Green) and turning inland on the west side of the churchyard. A turreted gatehouse, the Water Gate, allowed access onto the main thoroughfare, Castle Street. Little if anything remains of the walls today although the street plan is essentially the same as it was in the 17th century (Speed, 1611a). The development of Llangefni as the regional civic centre halted further residential

---

6 Llangefni, in central Anglesey, is a fine example of an 18th century market town but lies outside the area of this study.

and administrative expansion of Beaumaris, and today it stands as a well-preserved, Georgian spa and seaside resort, albeit one with a courthouse (constructed c. 1614, all stonework now obscured by render and limewash) and a gaol.

After the castle, the oldest building in Beaumaris is St Mary's and St Nicholas's Church, an elaborate 15th century structure in the Tudor style with large, tracery windows and prominent, decorative 'knobbly finials', this is a renovation of a 14th century building (Haslam et al., 2009). Certainly much of the stone used appears to have been reused at least once. Although built of coursed-rubble masonry, a number of the stones are incongruously well cut and dressed and look out of place in the fabric of the church. It is possible they could have been derived from decommissioned friary buildings in old Llanfaes or from former stretches of the town walls which may have enclosed the northern part of Beaumaris. What is certain is that the stone used is local. The bulk of the building is constructed from a mosaic of Penmon Loggerheads Limestone, weathering pale grey and shaley, dark grey, rust-weathering Leete Limestone. The latter stone is used abundantly in the construction of Beaumaris Castle. Quoins and a significant amount of the rubble masonry are Basement Beds grits, representing various facies and seemingly collected from all areas of the outcrop. Coarse grits and quartz-pebble conglomerates typical of the local outcrops at Penmon and Lligwy Bay are of course common, but red and white 'peaches and cream' stones from Moel-y-Don and green and purple facies typical of the Bridges Formation are also present. A few blocks of local green-schist also make a showing. The roof was raised in 1825 by Bangor architect John Hall. This work is clearly seen on the north side of the church where a clearly change in masonry style and material is visible. Hall used coursed ashlar masonry of local Basement Beds grits and conglomerates to add extra height to the aisles. The merlions above may be later; their greenish yellow tinge and the development of tafonic weathering suggest that they are Gwespyr Sandstone from the Talacre area of North East Wales, the only non-local stone used in the fabric of the church. Also of non-local stone is the war memorial in English, Upper Jurassic Portland Stone which stands in the churchyard. The churchyard is also of interest from its 18th century Cambrian slate gravestones and tomb slabs.

The town of Beaumaris we see today was mostly built between the 17th and 19th centuries, Basement Beds grits and Penmon Loggerheads limestone are the most abundantly used building materials, not a surprise given the proximity of Beaumaris to the quarries. Over the historical period, the most influential family in the town were the Bulkeleys, who owned the quarries at Penmon and considerable amounts of agricultural acreage on Anglesey and the mainland. The Bulkeleys had become established in Beaumaris before 1450. The tomb of William Bulkeley (d. *c*. 1490) and his wife Ellen Gruffudd is in St Mary's and St Nicholas's Church and is constructed of English alabaster from the Triassic Tutbury Gypsum Beds of Derbyshire. The family seat at Baron Hill is located just to the west of the town. Once an impressive, Italianate mansion, it was destroyed by fire in the 1940s, and its ruins still stand, decaying in the woods but on private, inaccessible land. According to Haslam et al. (2009), it is built of the local schist. The Bulkeleys had for four centuries stood in Parliament representing several constituencies and boroughs in Anglesey and mainland North Wales. The first Viscount, Thomas Bulkeley (d. 1659) supported the Royalist cause during the Civil War, garrisoned Beaumaris Castle and rallied Anglesey behind him. Nevertheless, he lost the Castle to the Parliamentarians. Living more quietly until the Restoration of the Monarchy, the Bulkeley lineage remained unbroken until the early 19th century. The seventh Viscount Bulkeley (Thomas James Warren Bulkeley, 1752–1822) was influential in the development of the modern town of Beaumaris. The main administrative centre on Castle Street is a building that combined the town hall on the upper floor and an arcaded market hall on the ground floor. This structure was commissioned by Bulkeley and completed in 1785. It is built from yellow-weathered, Basement Beds gritstones quarried locally.

Dating from the latest 18th century is the Bull's Head Hotel and its stable yard at the rear. The Hotel is coated in white render and so the original stone cannot be seen. However, the stable yard behind on Rating Row has been restored and converted into a restaurant and retail space. There is an unusual archway with a brick chimney on top, the span of the arch carries a flue from a nearby building. The stable block is probably very typical of buildings constructed in the town in the 17th and 18th century, being built of rubble masonry of local stone; Basement Beds grits and sandstones as well as numerous blocks of greenschist from the metamorphic terrane. The arch is dressed with rough-cut blocks of Penmon Limestone, including one with a fine example of a fossil *Lithostrotion* species colonial coral.

The start of the 19th century brought Beaumaris into a period of temporary grandeur. Foreseeing the building of Telford's Suspension Bridge at Porthaethwy (later Menai Bridge) which was to be constructed in 1828, the seventh Viscount Bulkeley had built the road from there to Beaumaris in 1805 and determined that Beaumaris would not be eclipsed by the new trade and travellers' route to Ireland. However, he died in 1822 without issue and without seeing his plans come into fruition. He was succeeded by his nephew, the somewhat over-named Sir Richard Bulkeley Williams-Bulkeley (1801–1875). Williams-Bulkeley shared his uncle's intention of keeping Beaumaris well and truly on the map. He bought the Castle and cleared it of debris and undergrowth and created the public park in its grounds. Williams-Bulkeley and the town corporation also invested in a major building scheme for Beaumaris and employed the architect firm Hansom and Welch to put this into action. Joseph Aloysius Hansom (1803–1882) partnered with Edward Welch (1806–1888) were well-known and influential 19th century architects and designers.[7] Hansom went into partnership with Welch in the late 1820s and the work at Beaumaris was one of their earlier commissions. They were also the first (and sadly the last) architects to see the potential of Penmon Limestone as a construction stone. They were responsible for building a bathing establishment (of which now only a chimney remains), the town gaol (1828), a number of terraced town houses (c. 1830; Figure 5.9), the sweeping Regency Victoria[8] Terrace (1833) and the Bulkeley Hotel (1835; Figure 5.10). None of these grand buildings would look out of place in the popular spa towns of England such as Cheltenham and Bath, and all were built using ashlar masonry from fossiliferous Penmon Limestone (Loggerheads Limestone). Hansom and Welch went on to design and build Birmingham's neoclassical Town Hall, an immense civic statement in the style of a Corinthian Roman Temple. Completed in 1834, this structure was also built of Penmon Limestone, and it remains one of the very few buildings built of the stone outside North Wales (Siddall et al., 2016). Further works might have

7 In addition to many buildings, Hansom designed the predecessors of London's black taxicabs, the lightweight, horse-drawn Hansom Cab.
8 Named after Princess Victoria who visited Beaumaris in 1832. Victoria became Queen in 1837.

Figure 5.9 A town house on Church Street, Beaumaris designed by Hansom & Welch and built c. 1830. This building has survived being coated with harling and painted.

been commissioned, but sadly the Birmingham project bankrupted the partnership, and it was dissolved in 1833.

On The Green is a Gorsedd Circle, placed to commemorate the 1996 Eisteddfod which was held in Beaumaris. This is built from slabs of Leete Limestone with large productid brachiopod fossils. This structure, a modern stone circle, is as good a place as any to see this more shaley limestone facies. The quarries, located at Penmon, beyond the town to the north are now overgrown and/or inaccessible or have become the location for other industries such as fish farming.

En route to Beaumaris from the mainland, it is necessary to pass through the old ferry port of Porthaethwy, now Menai Bridge. This small port town has most building stone well-and-truly obscured by harling and whitewash. However, two 19th century buildings are of great geological interest, being built of the local Precambrian Schists. The Victoria Hotel in Menai Bridge was built in 1852 with the intention of cashing in on the travellers' route to Ireland. It is built from mortared rubble masonry which is predominantly made from glaucophane schists from the Anglesey Penmynydd Terrane.

Figure 5.10 The Bulkeley Hotel in Beaumaris, designed by Hansom & Welch and built of Loggerheads Limestone from Penmon. (a) The hotel's façade on Castle Street and (b) a large, solitary, fossil coral in the building stone (field of view is 25 cm).

It therefore joins the limited number of buildings globally that are built from blueschist. The quoins are nicely cut ashlars, now whitewashed, they are probably Penmon Limestone. Nearby at the north end of the Menai Suspension Bridge on Mona Road is the local parish church, St Mary's. Built by Henry Kennedy in 1858, the church sits on a small hill. Again dressed with Penmon Loggerheads Limestone, and a few pieces of what appears to be *spolia* made from Malltraeth-facies Basement Beds sandstones, the main building material is schist rubble. This is predominantly greenschist rather than blueschist, and mainly the augen-schist facies with abundant quartz veins that outcrops on The Swellies islands in the Menai Straits and on the nearby shore beneath the Suspension Bridge.

Also on Anglesey, St Nidan's church on the outskirts of Brynsiencyn is included here as the most accessible building constructed of the largely quarried out Moel-y-Don red sandstone. St Nidans' Church was built between 1839 and 1843 by the architect John Welch (Haslam et al, 2009) to replace the decayed Medieval chapel located on a *llan* in the woods on the Menai Straits shore (Figure 5.11). The building is visible for miles and is noticeable by its top-heavy tower; a further storey including a clock and battlemented turret was added in 1933. Built from coursed, ashlar masonry, this church is a good place to see the variation in the stones acquired from the now largely quarried out Basement Beds gritstones which outcrop on the Anglesey shore of the Menai Straits at

Figure 5.11 The Church of St Nidan in Brynsiencyn is built of the local, Basement Beds red sandstones outcropping on the Anglesey shore of the Menai Straits.

Moel-y-Don and Pwllfanogl. These stones show a range of colours and textures in these well-bedded and cross-bedded sandstones and sandy limestones. They range from dark red to liver-coloured, medium-grained sandstones which are at first glance easily mistaken for the Triassic sandstones quarried in Cheshire, to pale orange-pink conglomerates, packed with rounded pebbles of vein quartz. In the middle are distinctive red sandstones with laminae of strings of small quartz pebbles (<10 mm diameter). Carbonate rich sands with scattered fossil crinoid ossicles are also observed. This latter facies is seen in the remaining outcrop at Moel-y-Don. A few blocks of grey limestone with fossil shells of productid brachiopods are also used in the masonry fabric.

5.6.3 Caernarfon

The name Caernarfon comes from the Old Welsh *Y gaer yn Arfon*; 'a fort in Arfon'. Before the building of the castle, the town was mostly located on the higher ground to the east of Twt Hill in the area of the Roman fort of Segontium and the parish church of St Peblig. This area is now known as Hendre and is a largely residential area, cut by the River Seiont. Following the construction of Caernarfon Castle and the town walls, Edward I's new civic centre was located on the coast at the mouth of the river. It has remained so until the present day, with county administrative buildings and shopping areas located within the Town Walls. In the centuries following the construction of the castle, there was little expansion of the town beyond the town walls. The town grew when Caernarfon became an important slate port and railhead in the 19th century, with the town centre taking over the area of the former royal fishpond on what is now Pool Street and residential areas growing to the north and east. The Roman fort of Segontium, the Castle and town walls already having been considered, the oldest buildings are the churches of St Peblig at Llanbeblig, St Mary's which is built into the town walls and the isolated St Baglan's Church to the south of the town, across the River Seiont.

Llanbeblig is named from a local saint called Publicus of Publicius. Nothing is known about him except that we can assume from his name that he was a Roman. The fact that a church would be founded in his name almost a millennium after the departure of the Romans from North Wales suggests a lingering familiarity with

Figure 5.12 St Peblig's Church, Llanbeblig. Multiple buildings phases in different stones can be seen in the tower. The lower parts of the tower and the main church are constructed from Basement Beds sandstones. The upper portions of the tower are predominantly the Leete Limestone facies of the Clwyd Group.

the period of Roman occupation and perhaps even a sense of *Romanitas*. Surprisingly, no substantial evidence of a pre-13th century church exists on this site, although the current structure is probably built on the original *llan* (Figure 5.12). First impressions of the fabric of the church leave the visitor in no doubt that the present structure is the result of several phases of building, this is particularly obvious in the tower. The oldest part of the structure and the probable remains of the earliest stone-built church here is the 13th century south wall of the nave. The tower, with its distinctive and unusual crow-stepped merlions and most of the walls were, for the most part, constructed in the 14th and 15th centuries. The Vaynol Chapel in the NE corner was added in the later 16th century. Further remodelling and repairs took place in 1775, mainly in the area of the north transept. The interior was completely reworked by the architect R. G. Thomas of Menai Bridge in 1894. During this phase of rebuilding, he also replaced the east window, quite unnecessarily, according to Haslam et al. (2009).

Given the church's close proximity to the Roman fort of Segontium, it is surprising that no obvious relics of Roman-period masonry exist in the church. The only exception is an altar which was found in the footings of south wall of the nave in the late 19th century and has now been removed. However, it is hard not to believe that this is *spolia* from the Roman site used in the fabric. Certainly, a wide mixture of stones is used in the construction of the building, though very little has previously been written concerning the materials used. The church presents a patchwork of regional stones, showing a definite temporal variation in lithologies.

Recently, the main stones used in the construction have been summarised in a regional survey by Davies (2018a), who recorded the coarse-grained conglomerates of the Basement Beds and the finer, buff to yellow Malltraeth Sandstones as the predominant building materials. A number of facies of these stones are present here, including breccias and gritstones, as well as iron-rich, orange-coloured grits and sandstones as well as a less-common facies of these sandstones with oxidation and reduction layers of a sage green and lavender to red-purple colour. These stones would have been quarried at various locations on Anglesey and on the Vaynol Estate on the mainland shore of the Menai Straits. Malltraeth sandstones are used for quoins and other dressings, though the coarse grits are used perhaps unexpectedly, for tracery on the window of the 16th century Vaynol Chapel. Window tracery has been replaced in some cases in later restorations; red Cheshire Sandstone is used to repair windows on the south wall of the nave, and a single quoin of this stone occurs on the south-west corner of the tower. The west transept was rebuilt in 1775 (RCAHMW, 1960) and uses a distinctive rust-red stone. This occurs in the tower and elsewhere in the building, including on the north wall of the nave. This is a shaley limestone quarried at Penmon and Moelfre on Anglesey and derived from the peritidal to lagoonal facies of the Lower Leete Limestone Formation. It contains large, thick-walled *Daviesiella llangollensis* brachiopods, and (rare) bedding plane surfaces show ripple structures. The rust colour is restricted to the weathering crust which is thin and prone to flake off on some surfaces, revealing a grey brown fresh surface.

These phases of construction and their equivalent stones can also be seen in the church tower. The Belfry has windows with louvres made from thin slates, a typically local use of slates quarried from the Cambrian Slate Belt. A naively carved head is located on the corbel of the nave roof as it adjoins the tower above the front porch. Carved from Malltraeth Sandstone, this probably dates to the 15th century (Haslam et al., 2009).

Beneath Thomas's 1890s' east transept window, the wall has been rebuilt with squared blocks of a strikingly white-weathering, porcellanous limestone. This stone breaks with a conchoidal fracture, and this can be seen on the dressed surfaces. This is a member from the Upper Leete Limestone Formation quarried at Tandinas as well as other locations on Anglesey. This stone has been stained orange from the rusting of the metal grills placed on the window above. The window itself has mullions and tracery of a drab sandstone, most likely Upper Carboniferous Cefn Stone from Broughton near Wrexham.

The interior of the church was also much modified by Thomas, who ripped out the plaster and revealed the stone walls. The main item of geological note is an impressive tomb in the Vaynol Chapel with effigies of 16th century Caernarfon patrician Thomas Gruffydd (d. 1587) and his wife Margaret (d. 1589). This monument was dedicated by their son, John. A shield in the window jamb, inscribed WG 1593, is also of the same stone. These are constructed, as is particularly typical of British church monuments of this style from English Alabaster.

In the centre of Caernarfon, St Mary's Church is built into the Town Walls and is contemporary with their construction and uses the same materials on the form of Basement Beds sandstones and gritstones. Initially built as a garrison church for Caernarfon Castle, it was modified and restored in 1814, with a new façade constructed in red Plas Brereton Sandstone. Paving along the façade uses beach cobbles, with pebbles of white vein quartz marking the 1814 construction date.

St Baglan's Church is a small church located on the coast 4 km west of Caernarfon, in the community of Llanfaglan. However, the church stands isolated in a large field. The original footprint of the church would have been a single rectangular chapel built in the 12th–13th centuries. This was extended and a south chancel added to this building in the 16th–17th centuries; the two phases of building are clearly seen on the north-facing wall. The church has been little used since this time, and the interior contains exceptionally well-preserved and largely unmodified wooden pews and pulpit dating from this time.

The older parts of the building are constructed from Basement Beds gritstones. However, the stone used on the 16th–17th century phases is unexpected; although local or Anglesey-derived Basement Beds gritstone continues to be used, there is a substantial amount of the texturally distinctive granodiorite from the Penryn Bodeilas Pluton on the eastern Llyn Peninsula. This medium to coarse-grained rock has distinctive clots of ferromagnesian minerals (hornblende, clinopyroxene and magnetite) in a matrix of lath-shaped plagioclase feldspar. The window dressings and the lychgate probably date to the 18th century and are built from a pink sandstone with prominent Liesegang bands, of the variety named herein the Plas Brereton Sandstone (see below for further use of this stone). The roof is constructed of local Cambrian Heather and Spotted Slates.

Expansion of the town of Caernarfon and a proliferation of building in wood, stone and plaster began in the 16th century. However, there are few structures remaining from this period, perhaps only the foundations of buildings. From paintings of the town produced by John Buckler in the first decade of the 19th centuries, it is known that a fine and large 16th century town house, Plas Mawr, with a double front, courtyard and stepped gables occupied a large plot on Castle Street (Buckler, 1810). This house belonged to the Gruffudd family commemorated in St Peblig's Church. From the yellow ochre paints used to paint this and other Caernarfon street scenes in Buckler's folio, one can assume that Basement Beds gritstone was used as the main building material. A number of buildings within the town walls date from the 17th and 18th centuries. The Black Boy Inn is the most substantial of these, and typical in that it is coated in stucco revealing little evidence of building stone. The uneven surfaces of the stucco coating suggest rubble masonry construction. In the 17th century, the town had expanded eastwards from the town walls into the area beyond Porth Mawr and around the pond formed by the damned River Cadnant; the area around what is now Pool Street (Speed, 1611b). The pond remained until the mid-19th century. It is most recently shown on John Wood's map of Caernarfon dating from 1834 (Cadw, 2010) and was only drained and built over during construction of the railway in the 1860s.

Plas Bowman (on High Street) and Plas Llanwnda (on Castle Street) date from the earliest 19th century and are also both built of yellow, coarse Basement Beds gritstone. These buildings would have been constructed shortly after Buckler made his paintings, and this period was the start of a major phase of redevelopment. Sixteenth and seventeenth century buildings were demolished and the streets lined with neat terraces and shop fronts. The site of Plas Mawr on Palace Street became the new market hall, built from large ashlar blocks of Loggerheads Limestone, some blocks contain spectacular examples of fossil colonial corals, particularly common are *Lithostrotion* species.

Georgian Terraces on the south end of Castle Street and other buildings including the Albert Inn on Rhes Segontium and the 1814 façade of St Mary's Church in the NW corner of the town walls are built of pink Plas Brereton Sandstone which is found uniquely as a building stone in Caernarfon and the immediate environs of the town. This is a fine-grained, rose-pink sandstone with well-developed Liesegang bands concentrating hematite. This stone was undoubtedly quarried very locally and possibly in the environs of Coed Helen or along the Menai Straits shore near Port Waterloo. Stone field walls near Coed Helen are built of the same stone as is the lych gate of nearby St Baglan's Church (constructed 1722; described above). The origin of this stone is unknown, and it has not been observed outcropping *in situ*. Its use is very much local to Caernarfon, and it has been tentatively assigned to the Upper Carboniferous Warwickshire Group, which outcrops in a narrow strip on the Menai Straits shore west of Caernarfon, to the north of the Aber Dinlle Fault. This stone is also used for the construction of Plas Brereton house and at the nearby Port Waterloo which lies just to the NE of the town (Figure 5.13).

Caernarfon Conservative Club, on the corner of Stryd Fawr (High Street) and Market Street, was extended upwards in 1886. However, the ground floor retains the original arcading of the former town meat market (the date of this original building is unknown, but it is again probably early to mid-19th century, perhaps built as early as 1813 (Cadw, 2010). The stone used to construct the arcades is the Ordovician Allt Llwyd Formation gritstones which would have been quarried locally in the town, from the south and eastern side of Twthill. This unit forms the base of the Nant Ffrancon Group. It is also remarkable to note that this stone is used extensively in the construction of the north-facing wall of Caernarfon Castle (built c. 1308) and that the Conservative Club is the only other structure in town to be constructed using this stone; two buildings separated by a period of at least 500 years (Figure 5.14). Contemporary records relating to the construction of the castle

Figure 5.13 Terraces on Castle Street, Caernarfon, constructed of pink Plas Brereton Sandstone.

Figure 5.14 The Conservative Club in Caernarfon, the ground floor was formally the meat market and was built in the early 19th century of Allt Llwyd Sandstone.

refer to a quarry at the 'town-end' (*ad finem ville*) which may well
have been the source of this stone in the early 14th century, and this
unit outcrops today at the foot of Twthill. This stone may have been
quarried from Twthill, or it may possibly represent spolia from an
earlier building.

The 19th century brought prosperity to Caernarfon as the main
port from which slates quarried at Nantlle were shipped abroad. In
the Medieval period, the quay had been along the town walls facing
the mouth of the River Seiont, but later developments took better advantage of the two rivers which flowed through the town, the
Seiont and the (now culverted) River Cadnant. Both of these rivers
flowing into the Menai Straits furnished the town with good quality, sheltered, albeit tidal, moorings. The change in location of the
main ports occurred in the later 18th century. A Custom House was
built against the exterior of the town walls, next to the water gate in
c. 1735, but by 1852 this had become the Anglesey Arms Hotel, and
the main port had moved upriver to the Slate Quay or further along
the Straits to the mouth of the River Cadnant. This allowed for the
area along the exterior of the walls to be developed as a promenade,
an idea only fully realised as recently as 2010.

The River Cadnant flowed from Y Maes (Pool Street) along
what is now Glan Ucha and Bank Quay, the latter street name re-
flecting this road's prior function as a riverside wharf. There was
probably a port, or at least a jetty at the mouth of the River Cadnant from as early as the 16th century, and certainly a small quay,
the 'patent slip', was in existence until 1834, but almost all traces
of this have been destroyed by the construction of Doc Fictoria
(Victoria Dock, originally called the New Basin) which began in
1868 and was completed in 1875. This new port was the brainchild
of Sir Llewellyn Turner, Mayor of Caernarfon and Constable of
Caernarfon Castle and a man determined to keep Caernarfon on
the map. Still surviving from the early 17th and 18th century dock
is a dry dock which lay adjacent to the patent slip and a cluster of
warehouses. Partially restored and operating as offices and a chan-
dler's shop, these are constructed of rubble masonry which includes
a significant amount of imported ballast material. The building
stones are locally derived basement beds grits of a typical coarse-
grained, yellow and rust weathering facies as well as the pink, fine-
grained variety of this formation which appears to be unique to
Caernarfon. Representing non-local ballast are abundant blocks of
grey, Newry Granodiorite, tourmaline-bearing Cornish granite as

well as other intrusive igneous rocks, mainly granitoids but also a few dolerites of unknown origin. A few pieces of metal-smelting slag occur in these buildings too. The harbour itself is constructed from substantial blocks of Lower Carboniferous Loggerheads Limestone quarried from Anglesey. These are of interest for their fossil fauna, with large colonial corals as well as solitary rugose corals clearly visible on the surfaces. The engineer responsible for the construction of the harbour was Frederick Jackson.

Just above Doc Fictoria, on the corner of Bangor Road is the former Christ Church. Built in 1855 by the architect Anthony Salvin, it is constructed of rubble masonry of Penmaenmawr Granodiorite with dressings and mouldings of a particularly yellow-coloured facies of Cefn Stone from the Wrexham area. Also nearby on the road to Bangor and dating from the early 19th century is the Celtic Royal Hotel (formerly the Uxbridge Arms Hotel; Haslam, 2009). This coaching house was built of well-dressed ashlars of pale brown-coloured sandstone. This is probably imported from NE Wales and may be Upper Carboniferous Gwespyr Sandstone from the area around Talacre.

Many paintings and prints were made of the picturesque view of Caernarfon Castle and the River Seiont in the 18th and early 19th centuries, and all depict boats moored along the east bank of the Seiont. The Slate Quay, as this wharf became known, was probably under construction around 1803, and a substantial dock was in place by 1834, managed by the Caernarfon Harbour Trust. The Quay itself is made ground, constructed from rubble removed from levelling *Y Maes*, with the river embanked along the front of the wharf (Cadw, 2010). Importantly the Slate Quay could be accessed without the need for goods to pass through the town and therefore slate could be brought directly from the quarries via the horse-drawn Nantlle Railway (1828) which later linked with the regional, mechanised rail network. The horse-drawn, narrow-gauge railway was constructed by the engineer father-and-son team Robert and George Stephenson and ran for almost 15 km. A purpose-built station was built on the Slate Quay, now refurbished as the terminus for the Welsh Highland Railway. The most prominent building on the Slate Quay was not a private concession of a quarry owner (which is the case at Port Penrhyn and Port Dinorwic). The port served all the independently owned quarries in the Nantlle area. The length of the quay was compartmentalised, with zones belonging to various slate quarry companies.

Figure 5.15 The Union Iron Works headquarters of the Slate Quay of Caernarfon. The patterned roof is in multi-coloured Cambrian slates from Nantlle.

The quay was also used for coal and materials for metal working, copper and iron ore. The Victorian Gothic headquarters of the Union Ironworks (established 1844[9]), although coated in harling and currently in a severe state of disrepair, have a beautiful patterned roof in purple and sage green Nantlle slates (Figure 5.15). This building is surrounded by industrial buildings and assembly shops. Behind the Harbour Trust building are a cluster of 19th century warehouses, again almost derelict but hopefully being considered for regeneration (Cadw, 2010). These are surprisingly well built and were warehouses, display and shipping offices for the Nantlle slate quarries. Some of these buildings are good surviving examples of slate cladding on building exterior walls. Their original intention was of course to showcase the high quality of the Cambrian slates of Arfon. The largest remaining building on the Slate Quay is a warehouse built up against the break in slope directly below Y Maes. This well-built, four-storey building belonged to Morgan

---

9 The Union Ironworks provided iron components used in the construction of the Britannia Bridge as well as infrastructure used in the slate quarrying industry. The Ironworks went out of business in 1901 (Cadw, 2010).

Lloyd & Son, Wine and Spirit Merchants, and was constructed in c. 1840 of Basement Beds gritstones in coursed rubble masonry with lintels above the doors and windows of Gwespyr Sandstone. The Basement Beds are predominantly rusty red-stained conglomerates and sandstones, with sharp boundaries between these, marking the base of channel fill. Also present in the masonry is the pink sandstone peculiar to Caernarfon.

Also around this time a new town centre was developed outside the town walls. The main town square, Y Maes, was laid out and functioned as a marketplace and later a bus station. To the north Pool Street and Bangor Road, heading north along the coast became the main access roads into town. New shop parades and public houses were built along these thoroughfares, many shopfronts were clad in the Norwegian monzonite larvikite with its distinctive schillerescent, antiperthitic feldspars. This stone was widely exported from 1880 onwards. Banks were built on the main square, using Penmon Marble and Millstone Grit sandstones, the latter imported from the English Pennines. The area of Turf Square east of Porth Mawr was also a smaller hub. Formerly the area of the 'Oatmeal Market' (Speed, 1611c), this area also has a fine former bank building, clad in architectural ceramic tiles and with a foundation of Emerald Pearl Larvikite.

Within the walled town, a new administrative district began to take shape. The neo-Palladian Caernarfon County Court, with its colonnaded portico, was built in 1863 of the Penmon limestone marble facies. Behind the Court House are a series of Victorian and modern administrative buildings. Nineteenth century buildings are clad using a rubble masonry of a grey granite. A major building stone of the later 19th and earliest 20th century is this ship's ballast in the form of a grey granodiorite, brought into the slate quay by the slate ships. This stone was used in renovations to Caernarfon Castle in the 1870s and was much remarked upon (and disapproved of) by Greenly (1932) and subsequent authors who have studied the construction stones of Caernarfon Castle. What these authors did not note is that this grey granite, which is most likely sourced from Newry in the Republic of Ireland, is also widely used as a building stone in the rest of the town, which was also undergoing reconstruction at this time. This granodiorite is used to construct the former County Gaol (now council offices) and gaoler's accommodation on Shirehall Street, a complex designed and built by John Thomas in 1867–1869. This stone is used at the English Presbyterian

Church (1882) and the Royal Naval Reserve Battery building (1888) on Doc Fictoria. It is also used for a number of residential buildings in the Twthill residential area development of the 1870s and 1880s. Wicklow Granite, the main building stone of Dublin, also occurs as ballast. This is a more blue-grey granitic rock with needle-like mafic minerals. This stone is used for building the Masonic Hall (formerly the English Wesleyan Chapel, 1867).

The residential area to the south east of Twthill was extensively redeveloped mainly between 1870 and 1890, following the cholera epidemic of 1860 (Rhydderch-Dart, 2017). This probably built over any viable outcrops of Allt Llwyd gritstones on the South eastern side of Twthill and indeed likely quarry sites for this stone. The majority of the buildings in the Twthill area are coated with harling and original building materials are not visible. Certainly local Caernarfon brick was a major construction material and a few buildings, including St Mary's Church, are clad with polygonal blocks of igneous rocks, a combination of ballast and Trefor Granodiorite. There is little evidence of the underlying Ordovician Allt Llwyd Grits being used here.

The residential area north of Twthill was not fully developed until the early 20th century when a school and cottage hospital and rows of Edwardian villas were built. This is the only area in which there is widespread use of Twthill Granite as a building stone, used from garden walls, particularly on Warfield Road and England Road South above this. Ballast, mainly Newry Granodiorite and some more exotic stones including amphibolites are used for wall building on England Road North. Further redevelopment of this residential area continued until the late 20th century.

The early 21st century has brought a better appreciation of Caernarfon's built environment and history beyond the castle, providing more facilities for both residents and visitors. The harbour at Doc Victoria has been restored and redeveloped as a marina. New buildings here are clad with Moelfre Limestone. The Slate Quay along the Aber Seiont, just north of the Castle, is largely a car park, but it is being redeveloped at the time of writing with older buildings restored and with the new terminus of the Welsh Highland Railway as a centrepiece. The town square of Caernarfon, Y Maes, was resurfaced using Cambrian and Ordovician Welsh Slate and setts of Trefor Granodiorite in 2008. Pennant Sandstone has been used for new retail developments on Bangor Road.

5.6.4 Conwy

The main town and administrative buildings of Conwy remain to this day enclosed within the Edwardian town walls. Residential areas have expanded into Gyffin and Cadnant Park only in the later 19th and 20th centuries. Prior to this, the Conwy was confined to the area enclosed by the town walls and managed to accommodate residential, administrative and commercial premises, and there remains a significant amount of residential property within the walled town. The street plan remains largely unchanged from that laid out by Edward I's town planners in the late 13th century, but only a few buildings exist from the Medieval and Renaissance period. These are the church of St Mary's, Aberconwy House, a merchants' house and warehouse dated to 1420 and the 16th century town house of Plas Mawr. Aberconwy House and Plas Mawr are remarkable survivors and outstanding examples of buildings of this type.

St Mary's Church almost certainly stands on the footprint of the pre-conquest church belonging to the Cistercian Abbey of Aberconwy. This establishment was removed upriver to Maenan when Edward I built his castle and established the new town of Conwy. The church has had several phases of rebuilding. The nave was probably constructed in the 14th century with the tower added in the 15th and 16th centuries. Significant modifications were made in the 19th century. The prominent architect Sir George Gilbert Scott reinstated a clerestory with a row of quatrefoil windows and replaced windows and the roof from 1872. His son, John Oldrid Scott continued work on the church into the later 1870s. Further work was undertaken in 1920 under Harold Hughes.

St Mary's Church is constructed from two main types of local stone, Conwy Rhyolite and Conway Castle Grit; both materials were also used for the construction of the Town Walls. The buff to yellowish-coloured Rhyolite is derived from the Conwy Volcanic Formation and/or Capel Curig Volcanic Formation. Both lithologies outcrop in close proximity at Bodlondeb, directly west of the town and on Conwy Mountain. The Conwy Castle Grits are a dark-brown weathering sandstone member of the Upper Ordovician (Ashgill) Conwy Mudstone Formation. These stones are laid as coursed, rubble masonry. A few other one-off blocks of sandstone, gritstone and limestone occur, but these are rare inclusions and do not suggest a systematic or intentional sourcing of these stones. The west door and several of the window dressings are carved from Gloddaeth Purple Sandstone, a stone also familiar from Conwy Castle. This is a sandstone with variable grain size, but it is predominantly a course grained gritstone. It is striking that it displays prominent cross-bedding and is mottled in colour between a dark purple red and a pale yellow to ivory colour. It has a tendency to weather along bedding and cross-bedding, which makes these structures all the more prominent when observed on building stones. This sandstone was sourced on the Creuddyn Peninsula, south of Llandudno in Bodysgallen park. It is now almost completely quarried out, and it has only relatively recently been recognised as a local building stone. Its use on some buildings in the Conwy area has been mis-attributed to the Triassic Cheshire sandstones. Gloddaeth Sandstone is also used for quoins on the west front, on the tower and for buttresses. Above the west door is a block in pale-brown sandstone, carved with a foliate design believed to date from the early 13th century (Haslam et al., 2009). It is too high to be directly observed but appears to be of Basement Beds grits from the Lower Carboniferous Clwyd Limestone Group (Figure 5.16). Comparing the crispness of carving in this block with the stiff-leaf capitals at the doorway in Gloddaeth Sandstone, the

Figure 5.16 The West Door of St Mary's Church, Conwy. The main building stone is coursed rubble of Conwy Castle Grits. The doorway dressings are in Gloddaeth Sandstone. Just above the door is a small carving of foliage in sandstone.

latter are much more deeply weathered, and this and the colour indicate that a higher-quality freestone was used for the brown sandstone block. A three-light window on the north chancel of the church is dressed in Clwyd Group limestone, probably derived from the Great Orme or Colwyn Bay area. This is a white-weathering, nodular limestone with fossil shell and coral debris. The chancel window on south side in the perpendicular style is by Sir George Gilbert Scott and executed in Cefn Stone.

Returning to the town, with the exception of the Castle and the church, the oldest building still standing is Aberconwy House, dated by dendrochronology to c. 1420 (Figure 5.17). This was once a merchants' house and warehouse, but it is now owned by the National Trust and functions as a shop. The uppermost storey is jettied and now whitewashed, but the wooden bracing of this timber and plaster built structure can by just discerned. The basement and raised ground floor storey accessed by an external staircase are stone-built in roughly coursed rubble masonry.

The lintel of the basement window (at pavement level) at the corner with High Street is a large glacially derived boulder of gritty, volcaniclastic rock, grey but weathering pale yellow brown. Otherwise dressings, corbelling and some of the rubble masonry are Gloddaeth Purple Sandstone, this building again shows the

Figure 5.17 Aberconwy House. Built predominantly from Conwy Castle Grits, the door dressings and corbels beneath the projecting window are Gloddaeth Sandstone.

importance of this local stone. The remainder of the rubble masonry are blocks of Conwy Castle Grits and Conwy Rhyolite, the latter weathering orange. The roof today is of Ordovician grey slate but would originally have been thatched.

The majority of the remaining notable stone buildings are on High Street. The grandest of all is Plas Mawr, built by the locally influential Wynn Family in 1576, is an astonishing survivor of its kind, described by Haslam et al. (2009) as 'half palazzo and half manor house', the interior is justly famous for its well-preserved, decorative plaster work. It is stone-built, but today completely plastered and whitewashed. According to Turner (2008), the stone used is local; Conway Castle Grits and very probably as at Aberconwy House, Conwy Rhyolite. Window dressings are of a red sandstone, which again according to Turner was sourced in Deganwy, though this does appear to be too fine-grained and uniform to be Gloddaeth Sandstone. Although constructed in the 19th century, the Castle Hotel on High Street was clearly influenced by the nearby Plas Mawr and was built in a Neo-Jacobean style in 1885, designed by the architect firm Douglas & Fordham. The façade with its shaped gables is dressed in brick and moulded architectural ceramics. Otherwise the façade is clad in an unusual style with fist-sized, angular blocks of Penmaenmawr Granodiorite, set into wet concrete. The porch and bow-window above are built from red Cheshire Sandstone. Also on High Street, late 19th century shops, such as the famous Edward's butcher's shop, are built from Clwyd Limestone and dressed with Cefn Stone. The same combination is used at the former bank (now a bar) on the corner of Rose Hill Street.

The Palace Cinema, which now houses shops and cafés, was designed by architect S. Colwyn Foulkes and built in 1935. The building is a masterpiece, somehow managing to project both *art deco* glamour whilst remaining in keeping with the town architecture. It would not be out of place on the set of an Errol Flynn movie. The façade is clad in random ashlar, with unusually shaped, polygonal quoins. Blocks are dressed to preserve chisel marks to give a slightly rusticated appearance. The façade is clad in mottled red and white Triassic Helsby Sandstone from Cheshire, which although uniformly fine-grained is a good substitute for the local Gloddaeth Stone. The same stone is used for the stepped, *art deco* façade on Drew Pritchard's antique shop at number 9 High Street (next door to the castle hotel), a structure probably of a similar date (Figure 5.18).

Figure 5.18 The façade of the Palace Cinema in Conwy is clad in mottled Cheshire Sandstone.

The area at the lower end of Rose Hill Street, opposite the Castle, is where the main administrative buildings of Conwy are located. The Guildhall, close to the Castle, was built in 1863 with some modifications made in 1925, in a mock-Tudor style. It is built from well-dressed random ashlar masonry of Conwy Castle grits. Many blocks contain pebbles and strings of conglomerate. The door and window dressings, including the word 'Guildhall' carved in gothic script, are in drab-coloured Cefn Stone. Additional decorative dressings around the north door utilise knapped blocks of Penmaenmawr Granodiorite. The roof is a nice example of the use of different coloured and shaped slate to produce a pattern. Cambrian purple and sage-coloured slates are used. Next door the town visitor centre is also in Conwy Castle Grits, as is the St John's Church (1881), opposite, this building, like the Guildhall is also dressed with Cefn Stone. Across the road from the Guildhall, on Castle Street the former Municipal Offices have a projecting ground floor shop front clad with local Clwyd Limestone. The façade of the upper storeys is a more modern use of Conwy Rhyolite, again with dressings and inscriptions in Cefn Stone. The visitor centre for the Castle, next door to the Guildhall on Llanrwst Road, is a new build clad in Pennant Blue Sandstone from South Wales.

The Town Quay, built along the estuary-facing line of town walls, was constructed in 1833 by one of Thomas Telford's engineers, William Provis, of Penmon Limestone, the same material used in Telford's Bridge crossing the Conwy Estuary. Buildings on the quay include a pub and a terrace of harling clad or white-washed cottages, including Britain's 'Smallest House' at no 10, with bright red painted harling.

St Mary's Church at Caerhun, situated some 7 km south of Conwy on the site of the Roman fort of Canovium in the Conwy Valley, is worthy of mention. The rectangular earthworks which demark the perimeter of the fort are still visible as a break of slope, and the St Mary's and its churchyard are situated in the NE quadrant of the fort's footprint. No *in situ* remains of Roman masonry are in evidence, but it would be difficult to believe that stone used in the Roman period was not used for the construction of the 14th–15th century Church.[10] Architecturally, St Mary's Church is a nave with a chancel projecting to the east. It is built of rubble masonry. A prominent and well-dressed ashlar block of red sandstone situated in the surrounds of the door is a contender for earlier use. The quality of cutting and chisel-marked dressing of this block, coupled with the likelihood of it being Triassic Cheshire Sandstone, makes it stand out from the other building materials use and suggests that it may have belonged to the Roman fort, but equally it could represent 20th century repair. Otherwise the church's rubble masonry consists of gritstones, slates, volcanic and volcaniclastic rocks and a small amount of limestone. Clastic sedimentary rocks include a small amount of Lower Carboniferous Basement Beds and a significant amount of mottled red and ivory Gloddaeth Purple Sandstone brought upriver from the Creuddyn Peninsula on the east bank of the Conwy Estuary. There is a small amount of Conway Castle Gritstone, weathering and dark, warm brown and Silurian, grey sandstones from the Denbigh Grits on the eastern side of the Conwy Valley. Much more abundant are the glassy, black porphyries of the Dolgarrog Volcanic Formation. These are flecked with white, euhedral feldspar crystals, 3 mm in length. Associated clastic and hyaloclastic rocks from the Trefriw Tuff Formation are also common. Both these units are basaltic in overall composition and overly the Snowdon Volcanic Group locally in the Conwy Valley.

---

10 Arguably any Roman buildings here would have been predominantly constructed of timber but it is likely such buildings would have had stone foundations.

The majority of the church's windows are dressed with Gloddaeth Purple Sandstone. The windows in the chancel and the east wall are of Conway Mudstone Formation slatey shales quarried from the Trefriw area. Finely and clearly laminated, with traces of ripple cross-bedding and often with very prominent tool marks from dressing the stone, this grey-brown, partially iron-stained slate initially resembles timber at first glance. It is used in thin batons to frame the windows. This stone is from thin sandstone bands bedded in the Uppermost Ordovician Conwy Mudstone Formation.

The church was restored by Henry Kennedy in 1851. The roof, originally thatched, was replaced with slate from the Ordovician Slate Belt (Nant Ffrancon Formation). Described in Haslam et al. (2009), as a traditional style, the valleys of the roofline are of note, using small slates to bridge the guttering where the roofs join. A sundial with a modern dial and gnomon stands at the SW corner of the church. This is built from Basement Beds gritstone. It does not require too much of stretching of the imagination to believe that in an earlier life that this stone was once a Roman capital.

It is likely that much of this stone used for building the church could have been collected from fluvial–glacial deposits in the river valley, rather than requiring quarrying. However, the presence of Gloddaeth sandstone indicates that this stone was being more widely used than the immediate vicinity of the town of Conwy and would have been intentionally brought to the site.

5.6.5 Harlech

Harlech is by far the smallest of the towns covered in this study, half the size of Conwy and Beaumaris and a fifth of the size of Caernarfon. It lies south of the main area under study. Although dominated by the castle, the adjoining town is not much more than a village in the parish of Llandanwg. It is divided into two parts, with the town centre clustered around the castle rock, with a residential area to the south and the lower town, beneath the castle on the coastal plain on the hanging wall of the Mochras Fault. John Speed's map of Merionethshire illustrates Harlech as a hamlet consisting of the Castle, a church and fewer than thirty scattered houses (Speed, 1611c). Once a tiny village, Harlech, and especially the lower town, grew in the 1860s when it became developed by quarry owner and entrepreneur Samuel Holland as a seaside resort following the coming of the railway line from Porthmadog. By the beginning of the 20th century, the dunes on the coastal plain had been capitalised as golf links, and this further added to Harlech's attractions as a tourist town, albeit with a fairly well-heeled and artistically minded clientele. Apart from housing in the lower town and a new school, there has been relatively little development in the 20th century.

Almost all the buildings in the upper town are constructed from the local, green Cambrian Harlech Grits (Rhinog Formation), which is the underlying bedrock of the upper town. Masonry styles range from random rubble to coursed rubble and coursed rubble dressed with ashlar quoins and window arches, as at St David's-in-Seion Church (formally the nonconformist Capel Seion) and the market hall on Twthil. A short parade of shops on Stryd Fawr are built from Harlech Grits using ashlar masonry, one with an impressive window surround of yellow Cefn Stone from the Upper Carboniferous of NE Wales. Terraces of 18th century houses are built from coursed rubble masonry composed of squared of blocks of the greenish gritstone. Notable are the large lintel slabs seen in these buildings, some more than 3 m in length (Figure 5.19). The Castle Hotel, built by Samuel Holland Junior in 1876 to accommodate his tourists, sits in the Upper Town opposite the Castle. It is

Figure 5.19 Cottages in Harlech built from Cambrian Rhinog Sandstones. The thickness of the blocks represents bedding thickness. Note the long slabs of Rhinog Sandstones used as lintels for doors and windows.

now in part the visitor entrance and exhibition hall of the castle. It too, built in a rather baronial style, is constructed of Harlech Grits. Roofing in the town has utilised the grey, Ordovician Slates from Blaenau Ffestiniog. This stone is also used for paving in the recently restored Castle visitor centre. Also part of recent redevelopment of the area directly surrounding the Castle is paving on the access road Twtil, a switchback road which connects the Lower and Upper Towns. Around the entrance to the Castle grounds, this has been recently paved with riven flags of Blue Pennant sandstone, imported from South Wales. Split bedding parallels these reveal spectacular examples of ripples and other sedimentary structures.

The lower town on the coastal plain consists of brick built, harling-clad housing and 20th century constructions in concrete, such as the modern Theatr Ardudwy Harlech (1973), built in cast concrete with lower walls faced in Harlech Grits.

The parish church in Harlech is St Tanwg's, replacing the abandoned old St Tanwg's church in the dunes at Llandanwg south of Harlech (see Chapter 3). Built in 1841, it was designed by Thomas Jones of Chester (Haslam et al., 2009) and built from Harlech Grits with window dressings of yellow buff sandstone, probably Cefn Stone imported from the Wrexham area. It contains a 15th century font which has been brought from the old church at Llandanwg carved from a yellow to grey weathering, medium-grained sandstone with distinctive rusty, iron spots. This is almost certainly another example of the local, but now quarried out, Egryn Stone.

5.7   THE SLATE PORTS

5.7.1   Porthmadog

As the main town on the route between Caernarfon and Harlech, Porthmadog is included here for the sake of completion and for its importance as a slate port. Both Porthmadog and nearby Tremadoc were towns founded in the early 19th century by William Madocks (1773–1828), a wealthy English landowner who owned land on the Glaslyn Estuary at Traeth Mawr. With the intention of increasing agricultural yield on his estate, he built a barrier across the Glaslyn 'The Cob'. The geological effect of this structure met Madocks's needs, but it also forced the River Glaslyn westwards, serendipitously and fortuitously increasing the depth and size of the harbour at Porthmadog. The town then rapidly developed a shipbuilding industry and also functioned as a slate port for the Ordovician slates quarried in Blaenau Ffestiniog, with the slate first borough down river by boat and later by the Ffestiniog Railway. It is not surprising, therefore, that the Ordovician slates are the ones predominantly used for roofing and cladding building façades in the town. This is in contrast to the majority of the towns discussed in this book, which, situated in Arfon, Conwy and Anglesey, predominantly use the Cambrian slate as a roofing material.

The Cob is 1.3 km long causeway across the Glaslyn Estuary linking Penrhyndeudraeth and the quarry at Minffordd (Garth Quarry), with Porthmadog (Figure 5.20). Prior to the Cob, it was only possible to pick one's way across the shifting sands at low tide or use a shallow-draft boat at high tide. Madocks employed local engineer John Williams to build the Cob and the causeway was completed in 1811. It is built of local dolerite rubble, quarried at Minffordd on the Penrhyndeudraeth bank, and possibly also from Moel-y-Gest above Porthmadog as well as the local shales of the Dol-cyn-afon Formation. Early construction was fraught with difficulty on the shifting sands of the Estuary. A trackway was built on a wooden scaffold, from which the stone could be tipped onto the line of the causeway at low tide. Rush matting was

Figure 5.20 The Cob crossing the Glaslyn Estuary at Porthmadog.

used to help secure the stone in place and stopped it from being washed away at high tide (Skempton et al., 2002; Crag, 1986). The Cob has subsequently been enlarged several times to carry the railway line and more recently to add a footpath. Minffordd Dolerite has been used in all phases of construction. The inland wall is clad in Ordovician Slate from Blaenau-Ffestiniog. The access to Porthmadog provided by the Cob was instrumental in its economic growth. The town became a local railway junction and then became rich as a slate port during the mid-19th century boom in slate production.

The main building stone used in Porthmadog in the 19th century was the local sandstone quarried from the Dol-cyn-Afon Formation of the Mawddach Group. This is of Lower Ordovician, Tremadocian age. It is a grey, finely laminated, ripple cross-bedded sandstone, weathering a rusty, dark-brown colour. Quarries are located in the town on the coast directly south of the harbour in the direction of Borth-y-Gest, and this stone was also probably quarried along with dolerite at Moel-y-Gest. Large, slab-like blocks could be acquired, up to 2 m long, and these are commonly seen in the buildings around the harbour area. This stone is used to construct the harbour and surrounding quays as well as most buildings in the harbour area and beyond.

Porthmadog's other local stone is the dolerite (microgabbro) quarried at Minffordd and Moel-y-Gest. As discussed above, it was being quarried for the construction of the Cob in the early 19th century. It was also used in the 19th century for rubble masonry construction in the town. It is used for a Victorian Villa (now HSBC Bank), which has dressings in Portland Stone as well as another former bank (now a bar) on High Street which is dressed with Bath Stone. Other examples are the Chapel on Bank Place and Salem Chapel on the high Street, both dressed with Penmon Limestone. A peculiar architectural style seen frequently in Porthmadog was presumably born of making use of dolerite rubble too small for masonry. Fist-size chips of dolerite are pressed into concrete render. This technique is used to effect on the Royal Sportsman Hotel (Figure 5.21) in Porthmadog and for domestic architecture elsewhere in the town. This form of surface decoration is also encountered in Groeslon, but otherwise it is a very localised technique. However, it is worth noting the Castle Hotel in Conwy also uses the same decorative effect. It is not known if the same builders were employed there.

Figure 5.21 The Royal Sportsman Hotel in Porthmadog, with a façade clad with dolerite rubble.

Blaenau-Ffestiniog grey, Ordovician Slate from the Nant Ffrancon Group is used universally as a roofing material, and a number of buildings in Porthmadog are also clad with slates. A spectacular example is Bridge House at the west end of the Cob, opposite the Harbour Station. This building originally had the role of the toll house for crossing the Cob. However, it later became the offices of Parc & Croesor Slate Quarries Ltd. (working the Ordovician slates), and it was at this time clad to showcase these quarries slates (Gwyn, 2015). Modern buildings at the station are also built of slabs and blocks of Ffestiniog slate.

An unusual artefact of Porthmadog's industrial past is Cei Ballast (also known as Lewis's Island), an artificial island formed from the section of harbour where the incoming slate ships dumped stone ballast before entering the port. No systematic geological study of the rubble of the island has been made to date, but to do so would prove to be a most informative and interesting geological exercise.

5.7.2 Y Felinheli/Port Dinorwic

Port Dinorwic came into being, as the English name suggests, as the slate port for the Dinorwic Quarries in Llanberis. This name has now largely gone out of use and has reverted to the older Welsh name Y Felinheli, meaning 'the salt mill' referring to pre-slate industry on the shores of the Menai Straits. A former hamlet of the village was Aberpwll which is named in the Exchequer accounts for Caernarfon Castle as a source of stone. This could have referred to limestone but is more likely to have worked Bridges Formation Basement Beds gritstone. It is also possible that it could have referred to the red sandstone quarry at Moel-y-Don which lies directly opposite on the Anglesey shore of the Menai Straits. Felinheli is occasionally also referred to as Moel-y-Don in some documents, as there was a ferry across the Straits here, leading to more confusion with toponyms and quarry sites. The local limestone, quarried on the Vaynol Estate, was used for masonry and also used to make lime. A number of kilns were operating on the shoreline in the first quarter of the 19th century. The lime industry was responsible for the first port in the village (Jones, 1992). The Vaynol Estate was owned by the Assheton-Smith family who also owned the quarries at Dinorwic in Llanberis, working Cambrian Slate. They built a new port at Y Felinheli, and hence it acquired the name Port Dinorwic. Slate was brought down from the quarry on a horse-drawn railroad which was constructed in c. 1825. This was superseded in 1843 by steam locomotives on the Dinorwic Quarries Railway, which later came to be known as the Padarn Railway.

A small village strung out along the Caernarfon-Bangor Road, Y Felinheli is constructed of two main stone types; pink, iron-stained crinoidal limestone quarried from the Lower Carboniferous strata of the nearby Vaynol Estate and Newry Granodiorite (also seen in abundance in Caernarfon) which was brought into the port as ship's ballast. This stone is used for the terraced cottages strung out along the main thoroughfare. Originally, these were probably cottages owned by dock workers and also slate quarrymen. Many have front garden walls and gateposts decorated in a local vernacular style, which involves encrusting these structures in big chunks of vein quartz, brought down from the mountains of Snowdonia (Figure 5.22).

Y Felinheli also owns a fine war memorial in Penmon Loggerheads Limestone. The local pink Vaynol Limestone is best seen at the former inn, the Halfway House, at the north end of the main village.

Figure 5.22 Gate posts decorated with vein quartz were popular in the 19th century. The Cottages here on Terfyn Terrace are built from Newry Granodiorite, brought into the port as ballast.

5.8 A FINAL NOTE ON GRAVESTONES AND MEMORIALS

Gravestones are a good subject upon which to conclude this book. They also demonstrate a continuity in the use of stones in NW Wales from the 5th century to the present day. The phenomenon of Early Christian Period inscribed stones, many of which were grave markers or memorials, and their substrates has been described in Chapter 3. They utilised materials from local sandstones to more unexpected use of hard, igneous rocks and even exploited the distinctive columnar jointing of lithified ash-flow tuffs. Grave markers of the later historical period are almost universally made from slate, up until the latter part of the 19th century. Few churchyards in the region have gravestones dating before the early 18th century (apart from extant Early Christian stones as at Llanaelhearn). Eighteenth century grave markers tend to be rectangular slabs of slate, in the region in question, these are predominantly heather and blue slates from the Cambrian Llanberis Slate Formation, often displaying reduction bands and spots. In the Conwy Valley and of course in Porthmadog, grey Ordovician Slate from the Blaenau-Ffestiniog area prevails. Letter-cutting is generally well executed but primitive in style and very frequently in English. Nineteenth century graves were marked by more intricately carved slate slabs as well as slate chest tombs (Figure 5.23). No systematic study has been made of the folk art of Welsh gravestones or a comparison made of the hands of carvers, but a series of motifs are frequently repeated across the region. Graves commonly feature a weeping willow tree (sometimes interpreted as a palm tree) carved in relief on the upper portion of the stone, sometimes with a pair of clasped hands above this. Sailors' graves often have a carving of an anchor cast upon rocks. However, many graves just bear a simple inscription.

From the last decade of the 19th century and into the 20th century, a trend for grave markers cut from imported, decorative Aberdeenshire or Scandinavian granites spread across Great Britain, and NW Wales was not an exception, although the local tradition for slate gravestones still continued. Most churchyards and cemeteries in the region have at least one, impressive, granite memorial of this type. Pillars topped with urns or finials and decoratively

Figure 5.23 St Aelhearn's Church and graveyard at Llanaelhearn. The majority of the graves here are in grey Ordovician Slate from Blaenau Ffestiniog. The carved 'weeping willow' in the grave in the right foreground was a popular decorative motif.

carved and etched grave slabs were popular. Such an example of an unusual, brown-coloured facies of Norwegian larvikite is that of the Hughes family in Llanaelhearn churchyard, dating from 1912. St Mary's Church at Caerhun in the Conwy Valley has early 20th century grave markers of Aberdeenshire granites (Corennie, Peterhead and Cairngall Granites) as well as a pillar with finial of South African Rustenburg Gabbro from the Bushveld Complex; a popular stone from funerary monuments and a by-product of the platinum mining industry in the Pretoria district. The graveyard at St Baglan's Church near Caernarfon contains a number of graves of the local landed gentry, the Humphreys, Jones and Armstrong-Jones families. Those dating from the later 19th century are in Scandinavian imported granitoids; the larvikite and red Swedish Balmoral Granite, another popular material. The grave of Antony Armstrong-Jones, Lord Snowdon (d. 2017) is in grey, Ordovician slate from Blaenau Ffestiniog. A few cases of the use of English Cornish granite, popularly carved into Celtic crosses, also occur. One of these, commemorating the incumbents of Llangristiolus Church is such an example. Graves of limestone and marble are relatively uncommon. Nevertheless, a spectacular example of a memorial in Penmon Marble (Loggerheads Limestone) is that of the poet *Eben Fardd* (Ebenezer Thomas, 1802–1863) in St Beuno's churchyard in Clynnog-Fawr with carving of foliage well demonstrating this stones quality as a freestone.

A gravestone worthy of note is that of the geologist who contributed so much to the understanding of the geology of NW Wales. If gravestones make a fitting conclusion to this book, then the one belonging to the most influential geologist of the region should mark the point where this book ends. Edward Greenly published the first geological map of Anglesey (which at the time of writing is still in print) as well as defining the stratigraphy of Arfon. He also contributed to the study of building stones, being the first to conduct a survey of the stones used in the construction of Caernarfon and Beaumaris Castles. Greenly died at his Bangor home, Aethwy Bridge on 4 March 1951. He was cremated and his ashes were interred with those of his wife, Annie (d. 1927), at Llangristiolus Church on Anglesey (Clarkson, 2004). It would perhaps have been fitting to have a grave marker of local stone. Perhaps a slab of Anglesey blueschist. However, Edward and Annie Greenly's gravestone, in the then fashionable style of an open book, is made from an even more ancient stone. This is Uthammar Granite, also known as

Figure 5.24 The grave marker of Edward Greenly and his wife Annie in the graveyard of St Cristiolus's Church at Llangristiolus on Anglesey. The stone, carved in the form of an open book, is a Proterozoic pyterlite granitoid from Sweden.

Bon Accord Red, a very coarse-grained facies of the Figeholm Pluton, quarried on the Kalmar Coast of Sweden near Oskarshamm. Technically this granite is part of the anorogenic rapakivi suite of granites intruded into the Proterozoic Fennoscandian Shield, a pyterlite. It is 1.4 billion years old (Figure 5.24).

Chapter 6

The building stones
of North West Wales
A final word

This book has shown how stone has been used in North West (NW)
Wales in both monumental and vernacular architecture over the last
six millennia, from the use of roughly shaped, glacially eroded field
stones to quarrying of slate on an industrial scale. Roofing slate,
of very high quality, has made Snowdonia globally important regarding the production of building stone, and although this stone
is certainly also of local importance, and used in many, often im-
aginative, applications, many other building materials have equal
prominence in this landscape. A wide range of Precambrian metamorphic rocks outcrop on Anglesey and an almost complete Lower
Palaeozoic succession is found in North Wales and the adjacent
English county of Cheshire. Almost all aspects of this regional geology have been collected, mined or quarried at some point and used
for stone tools, a source of ore, for building projects and for engineering works. Although the slate quarries and their tips dominate
the landscape, hundreds of other quarries, working all available
lithologies, exist within the region. These quarries range from small
pits to much more extensive limestone and granite workings. North
Wales is globally renowned for its geoheritage as well as its archaeological and industrial heritage, from Neolithic passage tombs
and cromlechs to the supreme military architecture represented
by the Castles of Edward I and then in the 19th century, Thomas
Telford's groundbreaking tubular and suspension bridges and the
engineering works that accompanied the boom in slate quarrying.
It is rightly designated a World Heritage Site (WHS) encompassing
The Castles and Town Walls of King Edward in Gwynedd, and the
broader approach of this book includes the landscape and geology
of the Slate Industry of North Wales, a region that is nominated for

DOI: 10.1201/9781003002444-6

WHS status. The geology of this region underlies its economy, both historically and now.

Slate is the most famous building stone known from the region, and the Cambrian slates of the Llanberis Formation are Global Heritage Stones. Slate is used in the region for roofing but also for paving, building stone and the construction of fences as well as many other applications. For example, Welsh writing slates were exported globally in the 19th and early 20th centuries and used by millions of schoolchildren worldwide. This stone occurs rarely in the fabric of the Edwardian Castles; the original roofs do not survive, though contemporary records of employing slaters and women who were collecting moss (used for laying slate) certainly suggest that buildings constructed within the castle wards had slate roofs. The castles are predominantly built from very local stones sources, often quarried from the foundations of the castles themselves. Harlech is built from Cambrian sandstones of the Rhinog Formation, which is the bedrock on which it stands. Conwy is built from the Conwy Castle Grits, which is again the substrate of the Castle. Caernarfon here is an exception, not utilising the underlying tuffs and shales as building stones. Both Caernarfon and Beaumaris Castles use large amounts of Lower Carboniferous sandstones, described herein as the Basement Beds, in their structures. The Basement Beds sandstones were clearly the most important building stones used in the region up until the mid-19th century. These rather unpromising appearing sandstones outcrop underlying and interbedded with Clwyd Group limestones on both Anglesey and the mainland and were clearly far more valued as masonry blocks than the limestone itself, which was quarried mainly for lime production. Leete Limestone from the Clwyd Group is the main limestone used in the building of Beaumaris Castle. This is a dark grey, rusty-weathering shaley facies of the limestone and is arguably the least attractive variety to be used as a building stone. The south face of Caernarfon Castle, which faces the River Seiont and the southern entrance to the Menai Straits, is clad in Cefn Mawr Limestone, which has been well cut into blocks of varying groups of sizes. This stone, both in terms of facies and masonry style (all the other castles are constructed from rubble masonry rather than ashlar), is only seen at Caernarfon, and therefore its use appears to be unique for the region at the time of building.

There are still questions to be answered which are beyond the scope of this book and require further dedicated research. For example, was stone from the Roman Fort of Segontium reused at Caernarfon Castle? This is certainly a possibility. The presence of well-cut limestone in the fabric is a unique feature of Caernarfon Castle. Indeed, Greenly (1932) was convinced that the Roman Fort was 'quarried' for stone used to build the Castle; his evidence was essentially based on common sense in that the Cefn Mawr Limestone was used extensively at Caernarfon, but it is rarely seen at Beaumaris which is located only 5 km along the coast from Flagstaff Quarry at Penmon. Greenly therefore reasoned that this limestone could only have come from Segontium. Although this cannot be proved, there is evidence from looking at ashlar block sizes that this stone was indeed *spolia*. Greenly and Aubrey Strahan had consulted with Mortimer Wheeler on the stone used at Segontium. Strahan had also suggested that red Cheshire sandstone had been used at Segontium and the presence of this stone fuelled Greenly's argument that Segontium was the source of much of Caernarfon Castle's building stone, as again neither of these stones were in use at Beaumaris, which again would have had easy access via ship to the quarries of the River Dee in Chester. However, there are contemporary records for both Caernarfon and Conwy Castles which clearly state that stone was contracted and brought from Chester for building works in the 14th century. Cheshire Sandstone has not been observed in the extant remains at Segontium by the author. However, a dark-red, liver-coloured sandstone derived from the Basement Beds outcrops at Moel-y-Don on Anglesey and is used at the Roman Fort, and it is suspected that this stone was mistaken by Strahan for Cheshire sandstone. A small number of blocks of genuine Cheshire Sandstone ashlar are to be found in the nearby St Peblig's church at Llanpeblig. Strahan was unfamiliar with the red sandstone quarried at Moel-y-Don through no fault of his own, as these strata had been almost entirely quarried out. Therefore spoliation of the Roman Fort is not directly demonstrated by the archaeological evidence. That stone and materials were robbed from the fort is not in doubt and was normal practice everywhere. Having been abandoned by some thousand years prior to the building of the Castle, there had been plenty of time for removal of stone from Segontium to take place. Surely St Peblig's Church had utilised the fort as a source of stone. Leete Limestone (which is also used at Beaumaris Castle) is used in the fabric of St Peblig's Church, but there is relatively little evidence of Cefn Mawr or Loggerheads Limestone associated with the earliest building phases of the church which had taken place in the 13th century. It could be argued that all this stone had been taken for building Caernarfon Castle. Penmon Limestone is used in areas local to the quarries in the late Roman and early Christian periods at the Din Lligwy settlement and the nearby 12th century Lligwy Chapel. Both structures are constructed of Loggerheads Limestone rubble rather than cut ashlar. There is little evidence to suggest that the limestone was exported and used more widely in the region. Therefore the occurrence of limestone ashlar in Caernarfon Castle is unique both temporally and regionally. A cursory examination of the stone used in the south wall of the Castle immediately demonstrates that all blocks are not of uniform size, which would be a reasonable expectation if the stone has been quarried for the sole purpose of building Caernarfon Castle. Instead it is constructed of groups of stone with similar aspect ratios, ranging from blocks of various sizes to slabs and small setts. Some of this variation in block size can be seen in the masonry of the Eagle Tower (Figure 6.1). Blocks of similar dimensions are used together as a single course or area of stonework rather than being randomly mixed. Distribution of these stones strongly suggests that they were reused from pre-existing buildings.

Despite its properties as an excellent building stone, Clwyd Group Limestones, and particularly that from the Cefn Mawr and Loggerheads Formations quarried at Penmon in north east Anglesey, were only really exploited on a large scale for building in the early 19th century. Chosen by the architects Hansom and Welch for building Beaumaris, it briefly became a popular stone, used for buildings locally as well as for the construction of ports and the bridges crossing the Conwy Estuary and the Menai Straits, and it was exported for use in the city of Birmingham. It also attained some kudos as a decorative stone, cut and polished and used for chimney pieces and ecclesiastical fittings, such as fonts. Sadly with the demise of Hansom and Welch's firm in the late 1830s, Penmon stone failed to reach a wider market as a stone for civic and domestic building. It was used extensively in engineering works, for the harbour at Port Penrhyn and marine architecture in Liverpool and on the River Mersey as well as for building Telford and Stephenson's bridges and the Marquess of Anglesey's Column.

Although shales and slates are abundant in outcrop throughout NW Wales, coarser-grained, quartz-rich sandstones are less common, at least in outcrop, and yet where they do (or did) outcrop, they were enthusiastically exploited as building stones. A number

Figure 6.1 The masonry of part of the Eagle Tower. The main building stone observed here is grey Cefn Mawr Limestone. A
stripe of Basement Beds sandstones can be seen at the top
of the photograph and across the middle of the photograph
at the top of the arrow loop and is also used for dressing the
loop. The use of the tafonic-weathering Gwespyr sandstone
is probably 19th century restoration. Note the groups of
different stone sizes in the limestone. There are uniformly
sized slabs as well as other groups of blocks of uniform dimensions, rather than a complete range of shapes and sizes.

of red and red and white mottled sandstones are used in the region,
and these lithologies are difficult to differentiate when seen out of
their geological context, and this has caused much confusion and
a history of misattribution of source quarries, as discussed above.
Red and mottled sandstones were quarried on Franciscan-owned
land at Moel-y-Don on the Anglesey shores of the Menai Straits,
opposite Port Dinorwic. These stones are better known from their
occurrence in buildings rather than *in situ*, as the stone is almost
entirely quarried out in this locality, leaving only scant outcrops
along the shoreline of the Moel-y-Don Peninsula on the Menai
Straits. Examining buildings near to this quarry source brings
about the most secure connection of source to site. The 16th century manor house of Plas Goch at Moel-y-Don and its surrounding
walls and the nearby church of St Nidan's at Brynsiencyn are useful structures for examining this stone. It seems for the most part,
this is a relatively fine-grained sandstone ranging in colour from
liver-coloured to red and red and white mottled members.

Gloddaeth Purple Sandstone like that from Moel-y-Don has also been largely quarried out and very small quantities of this stone remain in outcrop. However, the small exposures around Bodysgallen and Bryn Pydew on the Creuddyn Peninsula show it to be a very coarse-grained sandstone, in many places technically a gritstone, well cross-bedded and dramatically coloured buff, red and purple. It has been used in Conwy Castle as well as other medieval structures in the Conwy Valley region. This stone's occurrence in the fabric of St Mary's Church located on the Roman Fort at Caerhun may tentatively indicate that it was used in the construction of the fort itself. However, the navigable River Conwy could have also provided easy access to this stone in the 15th century at the time of the construction of the church. Nevertheless a survey of the use of this stone in the Conwy Valley has found it to be more extensively used than was previously thought.

Finer-grained, higher-quality freestones are obtained from the Permo-Triassic Sandstones of Chester. These too include mottled red and buff variety which are very similar in appearance to the stone from Moel-y-Don and have led to difficulty in assigning source quarries. The Cheshire region had been an important source of stone from the Roman period onwards, and it is entirely feasible that these stones could have been imported into North Wales. However, the Cheshire red sandstones tend to have a different colour to those found at Moel-y-Don and on the Creuddyn, being a brick red colour rather than liver-coloured or purple red. These stones were used at both Caernarfon and Conwy Castles for repairs subsequent to the original building phase.

Slate from NW Wales has roofed the world and justifiably deserves its GHS designation. However, new and forgotten building stones have been rediscovered and identified in fieldwork for preparation of this book. Pink and red sandstones with distinctive Liesegang banding apparently occur in association with a series of sedimentary rocks assigned to the Warwickshire Group on the mainland shores of the Menai Straits at Caernarfon. This stone is widely used in 18th to early 19th century domestic and ecclesiastical architecture in and around the town of Caernarfon. This stone is named here the Plas Brereton Sandstone after the 19th century house of this name and constructed of this stone. No outcrops of this stone were identified, but outside the civic centre of Caernarfon, clusters of use of this stone are found around Plas Brereton House and Port Waterloo to the north of Caernarfon and at Coed

Helen and Llanfaglan to the south. This is a newly characterised building stone for this region, and it seems that the sources had been quarried out by the mid-19th century at the latest.

New field work, also published here for the first time, has shown that Allt Lwyd Grits were derived from the town of Caernarfon and outcrop on the southern and eastern flanks of the small granite stock which forms Twthill and the use of this stone in 19th century buildings such as the former meat market (Caernarfon Conservative Club) in Caernarfon. This stone is well known from the north wall of Caernarfon Castle, constructed in the early 14th century, and Edward Greenly and others have ascribed the quarry *ad finem ville* (at Town-end), referred to in contemporary Exchequer Accounts, to be located on Twthill. However, this sandstone was assumed to have been worked out during the first two phases of construction of Caernarfon Castle. New outcrops have been revealed during late 20th century road construction, and it is possible that a small quarry may have existed on the south flank of Twthill in the early 19th century, in the area now occupied by a residential area built in the mid- to late 19th century.

Another 'lost stone', Egryn Stone was used in Welsh Medieval architecture and is listed in Exchequer Accounts for Harlech Castle and other contemporary records. Recently, Tim Palmer (Palmer, 2007) has secured the source of this stone near Barmouth and identified it in over twenty buildings in Merionethshire in addition to its use in Harlech Castle. Fieldwork conducted as research for this book has been able to extend the range of the use of Egryn Stone into Caernarfonshire and Gwynedd, where it is used in the dressings of St Beuno's Church at Clynnog-Fawr.

Other much forgotten stones are the igneous rocks, quarried in vast quantities from the granite and granodioritic stocks and plutons of Snowdonia and the NE Llyn Peninsula. Used to make paving setts and for railway ballast, these stones are now hidden beneath the asphalted streets of the Victorian towns and cities of the English Midlands and the North or beneath the tar and grime of the rail network. Easy to dismiss, these stones and their quarries leave behind a legacy of a major industry and, in some cases enormous, abandoned quarries. This industry is worthy of further research in terms of both social history and geoheritage.

Kings, princes, architects and engineers and even the slate quarry barons are well-known historic figures in the region of NW Wales and beyond; however, some lesser known characters who lived and worked in this landscape and were instrumental in understanding the geology and the development of the stone industry as well as the preservation of the castles deserve further recognition in the historical records. First and foremost is the geologist Edward Greenly, who applied his detailed knowledge of the local stratigraphy to building stones in the Edwardian Castles as well as to the Roman and Neolithic monuments of the region, working with eminent archaeologists including Mortimer Wheeler and Wilfred Hemp. Sir Llewellyn Turner was Constable of Caernarfon Castle in the late 19th century, and he was perhaps the first to fully grasp the importance of the Edwardian Castles as tourist attractions and crucial components of North Wales' cultural heritage. Turner's renovation and restoration of Caernarfon Castle employed the mason and builder John Jones, who will go down in history for his choice of the most inappropriate choice of stone which was nevertheless laid with immense skill within the Medieval walls. Samuel Holland junior is a name that crops up continually throughout this text. Holland established quarries in Blaenau Ffestiniog and was the main influence in the founding of the Ffestiniog Railway before turning his attention to the igneous rocks of the region during the 'granite bubble' which came in to being as the Victorian cities of England rapidly expanded. Holland also began to develop seaside resorts, transforming the prospects of Harlech having been constable of Harlech Castle in 1874 and active in local and national politics. Holland represented the Welsh Liberal Party as a Member of Parliament from 1870 to 1885.

The archaeology and history of NW Wales are intimately bound to its geology, which stands not only as a building stone resource for local and global construction and therefore its geology is important for the consideration of the current Anthropocene environment. This landscape and its geology have been a draw for tourists since the early 19th century, the region has been an important centre for rock climbing since the late 19th century, and there is continued development of geotourism through the Bronze Age copper mines of the Great Orme and the Slate Museums at Llanberis and Blaenau Ffestiniog. It is hoped that this book, with its focus on buildings stones and particularly those used in the Castles of Edward I, can contribute to this legacy by drawing attention to some of the forgotten and much disregarded stones of this historic region.

Gazetteer of sites mentioned in the text

Place	Grid Reference	Site
Aberconwy House	SH 781776	House
Alexandra Quarry	SH 517560	Slate Quarry
Arenig Quarry	SH 829391	Granite Quarry
Bachwen Cromlech	SH 407494	Burial Chamber
Bangor Cathedral	SH 580720	Church
Beaumaris Castle	SH 607762	Castle
Benllech Cromlech	SH 519826	Burial Chamber
Bodowyr Cromlech	SH 463683	Burial Chamber
Britannia Bridge	SH 541710	Bridge
Bryn Cader Faner	SH 648353	Circle Cairn
Bryn Celli Ddu	SH 507702	Passage Grave
Bryn-yr-Hen Bobl	SH 518690	Passage Grave
Bryncir Quarry	SH 481488	Sands & Gravel Quarry
Caer Leb	SH 472674	Roman Fort
Caer Nant (the Rivals) Quarry	SH 350450	Granite Quarry
Caerhun (Canovium)	SH 776703	Roman Fort
Caernarfon Castle	SH 477626	Castle
Caernarfon Slate Quay	SH 479625	Quay
Carreg-y-Llam	SH 335437	Granite Quarry
Castell Caer Seion (Conwy Mountain)	SH 760778	Hillfort
Cedryn	SH 719635	Slate Quarry
Cei Ballast (Lewis's Island)	SH 570379	Ballast dump
Chwarel Ddu	SH 721521	Slate Quarry
Cilgwyn Quarry	SH 501540	Slate Quarry
Clogwyn-y-Fwch	SH 759618	Slate Quarry
Coed Helen	SH 474622	Landmark
Conwy Castle	SH 783774	Castle
Conwy Castle Quarry	SH 777773	Sandstone Quarry

(*Continued*)

Place	Grid Reference	Site
Conwy Mountain Quarry	SH 754777	Rhyolite Quarry
Conwy Suspension Bridge	SH 784774	Bridge
Criccieth Castle	SH 499377	Castle
Cwm Eigiau	SH 708635	Slate Quarry
Cwt-y-Bugail	SH 734468	Slate Quarry
Diffwys	SH 712463	Slate Quarry
Din Lligwy	SH 497861	Romano-British Village
Dinas	SH 273359	Granite Quarry
Dinas Dinlle	SH 437563	Hillfort
Dinas Dinorwig	SH 549653	Hillfort
Dinas Gynfor	SH 390951	Hillfort
Dinorben Quarry	SH 58 82	Limestone Quarry
Dinorwig Quarry Complex	SH 596603	Slate Quarry
Doc Fictoria	SH 478630	Quay
Dol Ifan Githin Quarry	SH 540505	Slate Quarry
Dolbadarn Castle	SH 585598	Castle
Dolwyddelan Castle	SH 721523	Castle
Dorothea Quarry	SH 500532	Slate Quarry
Druid's Circle	SH 722746	Stone Circle
Egryn	SH 609206	Sandstone Quarry
Flagstaff Quarry	SH 635808	Limestone Quarry
Fort Belan	SH 440609	Fort
Gimblet Rock (Carreg-yr-Imbill)	SH 388346	Dolerite Quarry
Gloddaeth Quarry	SH 795800	Sandstone Quarry
Glyn Cromlech	SH 514817	Burial Chamber
Glynllifon Estate	SH 456553	House
Graig Lwyd	SH 701753	Granodiorite Quarry
Great Orme Copper Mines	SH 770831	Copper Mine
Groby (Ffestiniog) Quarry	SH 695452	Granite Quarry
Gwalchmai Quarry	SH 381765	Granite Quarry
Gwydyr Castle	SH 795610	House
Gwylwyr	SH 318413	Granite Quarry
Gyrn Ddu Quarry	SH 394 466	Granite Quarry
Hafod Las Quarry	SH 779562	Slate Quarry
Harlech Castle	SH 580312	Castle
Idwal Cottage	SH649603	Visitor Centre
Llanaber Church	SH 599180	Church
Llechwedd Slate Mines	SH 700470	Slate Quarry
Lligwy Chapel	SH 499863	Church
Lligwy Cromlech	SH 501860	Burial Chamber
Maenan Abbey	SH 789656	House
Maenan Quarry	SH 790657	Sandstone Quarry
Maenofferen	SH 714465	Slate Quarry
Malltraeth	SH 401687	Sandstone Quarry

(Continued)

Place	Grid Reference	Site
Manod	SH 727455	Slate Quarry
Manod Quarry	SH 708447	Dolerite Quarry
Marquess of Anglesey's Column	SH 534715	Monument
Menai Suspension Bridge	SH 556714	Bridge
Minfordd (Garth)	SH 594391	Dolerite Quarry
Moel Tryfan Quarry	SH 515559	Slate Quarry
Moel-y-Don Quarry	SH 518677	Slate Quarry
Moel-y-Gest Quarry	SH 555389	Dolerite Quarry
Moelfre Aber Quarry	SH 502865	Limestone Quarry
Nanhoron	SH 287329	Granite Quarry
Nant Newydd Quarry	SH 480812	Limestone Quarry
Nant Quarry	SH 352454	Granite Quarry
Nant-y-Carw	SH 236323	Granite Quarry
Nantlle Quarry Complex	SH 502533	Slate Quarry
Oakeley	SH 691470	Slate Quarry
Pant-y-Carw Quarry	SH 782622	Dolerite Quarry
Parciau	SH 494846	Hillfort
Pen-y-Gaer	SH 297283	Granite Quarry
Pen-yr-Orsedd Quarry	SH 509539	Slate Quarry
Penmon Priory Complex	SH 630807	Church, Well and Dovecote
Penmon Quarry 1	SH 630813	Limestone Quarry
Penmon Quarry 2	SH 635813	Limestone Quarry
Penrhyn Bodeilas	SH 318421	Granite Quarry
Penrhyn Castle	SH 602718	Castle
Penrhyn Quarry (Caebraichycafn)	SH 616644	Slate Quarry
Perthi Duon Cromlech	SH 479667	Burial Chamber
Plas Brereton	SH 491644	House and Estate
Plas Coch	SH 511684	House
Plas Glyn-y-Weddw	SH 328313	Art Gallery
Plas Mawr Conwy	SH 780776	House
Plas Newydd	SH 520695	House
Plas Newydd Cromlech	SH 519697	Burial Chamber
Plas Tan-y-Bwlch	SH 655406	House
Plas-yn-Penrhyn	SH 590379	House
Port Dinorwic	SH 523676	Quay
Port Penrhyn	SH 591730	Quay
Port Waterloo	SH 487642	Port
Porth-y-Nant	SH 349448	Village
Porthmadog Harbour	SH 570384	Quay
Rhuddlan Bach Quarry	SH 485808	Limestone Quarry
Segontium	SH 485624	Roman Fort
St Aelhearn's Church	SH 387448	Church
St Aelhearn's Well	SH 384446	Well House
St Baglan's Church	SH 455606	Church

(Continued)

Place	Grid Reference	Site
St Beuno's Church	SH 414496	Church
St Beuno's Well	SH 413494	Well House
St Cristiolus's Church	SH 450736	Church
St Mary and All Saints Church, Conwy	SH 781775	Church
St Mary's Church, Caerhun	SH 776703	Church
St Mary's Church, Caernarfon	SH 477629	Church
St Tanwg's Church, Harlech	SH 581310	Church
St Tanwg's Church, Llandanwg	SH 568282	Church
Tan-y-Graig	SH 394466	Granite Quarry
Tandinas Quarry	SH 583819	Limestone Quarry
The Cob	SH 577381	Coastal Barrier
Tomen-y-Mur	SH 705386	Roman Fort
Town End Quarry, Caernarfon	SH 482631	Sandstone & Granite Quarry
Tre'r Ceiri	SH 374447	Hillfort
Trefor	SH 362460	Granite Quarry
Tremadoc Bath House	SH 557401	Roman Bath
Ty Newydd Cromlech	SH 344738	Burial Chamber
Ty'r Mawn (Gwydyr) Quarry	SH 788610	Dolerite Quarry
Tyddyn Hywell	SH 398471	Granite Quarry
Vaynol Estate	SH 536694	Estate and Manor House
Vivian Quarry	SH 586605	Slate Quarry
Waunfawr Quarry Complex	SH 551600	Slate Quarry

References

Alcock, L., 1968, Excavations at Deganwy Castle, Caernarfonshire, 1961–6, *The Archaeological Journal*, 124, 191–201.

Allen, P. M. & Jackson, A. A., 1985a, *Geology of the Country Around Harlech*, Memoir of the British Geological Survey, British Geological Survey & Her Majesty's Stationary Office (HMSO), London, 112 pp.

Allen, P. M. & Jackson, A. A., 1985b, *Geological Excursions in the Harlech Dome*, Classical areas of British Geology, British Geological Survey & Her Majesty's Stationary Office (HMSO), London, 94 pp.

Asanuma, H., Fujisaki, W., Sato, T., Sakata, S., Sawaki, Y., Aoki, K., Okada, Y., Maruyama, S., Hirata, T., Itaya, T. & Windley, B. F., 2017, New isotopic age data constrain the depositional age and accretionary history of the Neoproterozoic-Ordovician Mona Complex (Anglesey-Lleyn, Wales), *Tectonophysics*, 706–707, 164–195.

Ashbee, J., 2006, The royal apartments in the inner ward at Conwy Castle, *Archaeologia Cambrensis*, 153, 51–72.

Ashbee, J., 2015, *Conwy Castle*, Cadw, Cardiff, 64 pp.

Ashbee, J., 2017a, *Harlech Castle*, Cadw, Cardiff, 44 pp.

Ashbee, J., 2017b, *Beaumaris Castle*, Cadw, Cardiff, 56 pp.

Ashdowne, R. K., Howlett, D. R. & Latham, R. E., 2018, *The Dictionary of Medieval Latin from British Sources*, The British Academy and Oxford University Press, Oxford, England. Online version, http://www.dmlbs.ox.ac.uk/web/online.html.

Avent, R., 2010a, *Dolwyddelan Castle, Dolbadarn Castle, Castel y Bere*, Cadw, Cardiff, 40 pp.

Avent, R., 2010b, The conservation and restoration of Caernarfon Castle 1845–1912, in: Williams, D. M. & Kenyon, J. R. (Eds.), *The Impact of the Edwardian Castles in Wales*, Oxbow Books, Oxford, United Kingdom, 140–149.

Avent, R., Suggett, R. & Longley, D., 2011, *Criccieth Castle, Penarth Fawr Medieval Hall-House, St Cybi's Well*, Cadw, Cardiff, 56 pp.

Baillie Reynolds, P. K., 1938, Excavations on the site of the Roman fort of Kanovium at Caerhun, Caernarvonshire: Collected reports on the excavations of the years 1926–1929 and on the pottery and other objects found, Kanovium Excavation Committee. Cardiff, William Lewis, Printers, 282 pp.

Baines, J. A., Newman, V. G., Hannah, I. W., Douglas, T. H., Carlyle, W. J., Jones, I. L., Eaton, D. M., Zeronian, G. & CEGB, 1983, Dinorwig pumped storage scheme. Part 1: Design. Part 2: Construction, *Proceedings of the Institution of Civil Engineers,* 74(4), 637–718.

Ball, T. K. & Merriman, R. J., 1989. The petrology and geochemistry of Ordovician Llewelyn Volcanic Group, Snowdonia, North Wales. British Geological Survey Research Report, SG/89/1, Keyworth, Nottingham, 23 pp.

Barber, A. J. & Max, M. D., 1979, A new look at the Mona Complex (Anglesey, North Wales), *Journal of the Geological Society,* London, 136, 407–432.

Barnard, P. & Collins, A., 2007, Geoscience Wales Limited Field Course for Libyan Petroleum Institute, Applied Petroleum Technology (UK) Limited, Colwyn Bay, North Wales, 21 pp.

Barnwell, E. L., 1883, Dolwyddelan Castle, *Archaeologia Cambrensis,* 38, 51–56.

Barrett, T. J., MacLean, W. H. & Tennant, S. C., 2001, Volcanic sequence and alteration at the Parys Mountain volcanic-hosted massive sulfide deposit, Wales, United Kingdom: Applications of immobile element lithogeochemistry, *Economic Geology,* 96, 1279–1305.

Bates, D. E. B. & Davies, J. R., 1981, *Geologists' Association Guide no. 40: Anglesey,* Geologists' Association, London.

Baynes, E. N., 1908, The excavations at Din Lligwy, *Archaeologia Cambrensis,* 6th Series, 8, 183–210.

Bendall, C., 2004, Trefor Granite, *Welsh Stone Forum Newsletter,* 2, 2.

Bergström, S. M., Chen, X., Gutiérrez-Marco, J. C. & Dronov, A., 2009, The new chrono-stratigraphic classification of the Ordovician System and its relations to major regional series and stages and to δ^{13} C chemostratigraphy, *Lethaia,* 42, 97–107.

Bonney, T. G. & Houghton, F. T. S., 1879, On the metamorphic series between Twt Hill (Caernarvon) and Port Dinorwig, *Quarterly Journal of the Geological Society,* 35, 321–326.

Boon, G. C., 1960, A temple of Mithras at Caernarvon Segontium, *Archaeologia Cambrensis,* 109, 136–172.

Borradaile, G. J., MacKenzie, A. & Jensen, E., 1991, A study of colour-changes in purple-green slate by petrological and rock-magnetic methods, *Tectonophysics,* 200, 157–172.

Brasier, A. T., Rogerson, M. R., Mercedes-Martin, R., Vonhof, H. B. & Reijmer, J. J. G., 2015, A test of the biogenicity criteria established for microfossils and stromatolites on quaternary tufa and speleothem materials formed in the "Twilight zone" at Caerwys, UK, *Astrobiology,* 15(10), 883–900.

Brewer, J. N., 1825, *Delineations of Gloucestershire*, Republished by The History Press (2005), London, 202 pp.

British Geological Survey, 1852, Sheet 78, Anglesey, South of Holyhead Island, Beaumaris, Geological Survey of England and Wales 1:63,360 geological map series [Old Series], British Geological Survey.

British Geological Survey, 1980, Anglesey sheets 92 & 93 and parts of sheets 94,105 & 106 (special sheet), Geological Survey of England and Wales 1:50,000 geological map series, New Series, British Geological Survey, Keyworth.

British Geological Survey, 1982, 135 & part of 149, Harlech and part of Barmouth, Geological Survey of England and Wales 1:50,000 geological map series, new series, British Geological Survey, Keyworth.

British Geological Survey, 1985a, 106, Bangor (Solid), Geological Survey of England and Wales 1:50,000 geological map series, new series, British Geological Survey, Keyworth.

British Geological Survey, 1985b, 34, sheet SH 65 and SH 66 (parts of sheets) Passes of Nant Ffrancon and Llanberis, 1:25 000 geological map series: Classical areas of British geology, British Geological Survey, Keyworth.

British Geological Survey, 1986, 35, sheet SH 66 and SH 67 (parts of sheets) Bethesda and Foel-Fras, 1:25 000 geological map series: Classical areas of British geology, British Geological Survey, Keyworth.

British Geological Survey, 1989a, 94, Llandudno; (solid & drift), Geological Survey of England and Wales 1:50,000 geological map series, new series, British Geological Survey, Keyworth.

British Geological Survey, 1989b, 41, sheet SH 77/78 Conwy, 1:25 000 geological map series: Classical areas of British geology, British Geological Survey, Keyworth.

British Geological Survey, 1997, 119, Snowdon (solid & drift), Geological Survey of England and Wales 1:50,000 geological map series, new series, British Geological Survey, Keyworth.

British Geological Survey, 2015, 118, Nefyn and part of Caernarfon (solid & drift), Geological Survey of England and Wales 1:50,000 geological map series, new series, British Geological Survey, Keyworth.

Bromley, A. V., 1965, Intrusive quartz-latites in the Blaenau Ffestiniog area. *Geological Journal*, 4(2), 247–256.

Buckler, J., 1810, Ecclesiastical, monumental and castellated antiquities of North Wales, F. T. Sabin, 78 pp.

Burek, C. V., 2008, History of RIGS in Wales: An example of successful cooperation for geoconservation, in: Burek, C. V. & Prosser, C. D. (Eds.), *The History of Geoconservation*, Geological Society, London, Special Publications, 300, 147–171.

Burl, A., 2000, Chapter 10: The stone circles of Wales, in: *The Stone Circles of Britain, Ireland and Brittany*, Yale University Press, New Haven and London, 175–193.

Butler, L., 2010, The castles of the princes of Gwynedd, in: Williams, D. M. & Kenyon, J. R. (Eds.), *The Impact of the Edwardian Castles in Wales*, Oxbow Books, Oxford, United Kingdom, 27–36.

Butler, L. A. S. & Evans, D. H., 1980, The Cistercian Abbey of Aberconway at Maenan, Gwynedd: Excavations in 1968, *Archaeologia Cambrensis*, 129, 37–63.

Cadw, 2010, *Caernarfon Waterfront: Understanding Urban Character*, Cadw, Parc Nantgarw, Cardiff, 55 pp.

Cadw, 2018, *World Heritage Site Management Plan 2018–28, Castles and Town Walls of King Edward in Gwynedd*, Welsh Government Historic Environment Service (Cadw), Cardiff, 193 pp.

Caffell, G., 1983, *Llechi Cerfiedig Dyffryn Ogwen/The Carved Slates of Dyffryn Ogwen*, National Museum of Wales, Cardiff, 24 pp.

Cameron, D. G., Evans, E. J., Idoine, N., Mankelow, J., Parry, S. F., Patton, M. A. G. & Hill, A., 2020. *Directory of Mines and Quarries*, 11th Edition, British Geological Survey, Keyworth, Nottingham.

Campbell, S., Wood, M. & Windley, B. F., 2014, *Footsteps Through Time: The Rocks and Landscape of Anglesey Explained*, GeoMôn, Isle of Anglesey County Council, 193 pp.

Casey, P. J., Davies, J. L. & Evans, J., 1993, Excavations at Segontium (Caernarfon) Roman fort, 1975–1979, CBA Research Report 90, Council for British Archaeology, London, 346 pp.

Cattermole, P. J. & Romano, M., 1981, *Geologists Association Guide: Lleyn Peninsula No. 39*, Geologists' Association, London, 39 pp.

Chapman, E. M., Hunter, F., Booth, P., Wilson, P., Pearce, J., Worrell, S. & Tomlin, R. S. O., 2013, Roman Britain in 2012, *Britannia*, 44, 277–343.

Clark, E., 1850, *The Britannia and Conway Tubular Bridges with General Enquiries on Beams and on the Properties of Materials Used in Their Construction*, John Weale & Day and Son, London, 466 pp.

Clarkson, E. N. K., 2004, Edward Greenly 1865–1951, Oxford Dictionary of National Biography, doi: 10.1093/ref:odnb/37485.

Clough, T. H. McK. & Cummins, W. A. (Eds.), 1988, Stone Axe Studies II—The petrology of prehistoric stone implements from the British Isles. CBA Research Report No 67.

Coldstream, N., 2003, Architects, advisers and design at Edward I's Castles in Wales, *Architectural History*, 46, 19–36.

Coldstream, N., 2010, James of St George, in: Williams, D. M. & Kenyon, J. R. (Eds), *The Impact of the Edwardian Castles in Wales*, Oxbow Books, Oxford, United Kingdom, 37–45.

Cragg, R., 1986, Chapter 1: North Wales, in: *Civil Engineering Heritage Wales and West Central England*, ICE Publishing, London, 5–53.

Crew, P., 1990, Late iron age and Roman iron production in North-West Wales, in: Burnham, B. C. & Davies, J. L. (Eds.), *Conquest, Co-existence* and Change: Recent Work in Roman Wales, Trivium 25 (Lampeter, Wales), St David's University College, Carmarthen, Wales, 150–160.

Cunliffe, B., 2001, Facing the Ocean: The Atlantic and Its Peoples, Oxford University Press, Oxford, 600 pp.

Dallmeyer, D. & Gibbons, W., 1987. The age of blueschist metamorphism in Anglesey, North Wales: Evidence from $^{40}Ar/^{39}Ar$ mineral dates of the Penmynydd Schists, Journal of the Geological Society, London, 144, 843–852.

Daniell, W. & Ayton, R., 1814, Black marble quarry near Red Wharf Bay, Anglesea, A Voyage Around Great Britain, Longman, Hurst, Rees, Orme, and Brown; and William Daniell, London, 307 plates.

Davies, I. E., 1974, A history of the Penmaemawr quarries, Transactions of the Caernarvonshire Historical Society, 35, 27–72.

Davies, D. C., 1880, A Treatise on Slate and Slate Quarrying, Crosby, Lockwood & Co., London, 186 pp.

Davies, J. R., 1982, Stratigraphy, sedimentology, and palaeontology of the lower carboniferous of Anglesey, Keele University, Unpublished Ph.D. Thesis, 270 pp + 79 plates.

Davies, J. R., 1983, Stratigraphy, sedimentology and palaeontology of the lower carboniferous of Anglesey, Unpublished Ph.D. Thesis, University of Keele, 270 pp + 79 plates.

Davies, J., 1993, A History of Wales, Penguin Books, London, 718 pp.

Davies, J., 2011, Building stone reports: No.1; St Tanwg's Church, Llandanwg, Meirionydd, [SH 5728], Welsh Stone Forum Newsletter, 8, 25–26.

Davies, J. R., 2003, The use of Carboniferous sandstones and grits from Arfon and Môn as a freestone over a wide area from Llandudno to northern Ceredigion, Welsh Stone Forum Newsletter, 1, 6.

Davies, J. R., 2016, Building stones in churches across Wales: A national map of vernaculars: Part 3. Ynys Môn (Anglesey) & Merionedd, Welsh Stone Forum Newsletter, 13, 2–10.

Davies, J., 2018a, Building stones in churches across Wales: A national map of vernaculars: Part 5: Caernarfonshire, Welsh Stone Forum Newsletter, 15, 7–11.

Davies, S. (translator), 2018b, The Mabinogion, Oxford University Press, Oxford, United Kingdom, 336 pp.

Davies, J. R., Somerville, I. D., Waters, C. N. & Jones, N. S., 2011, Chapter 8: North Wales, in: Waters, C. N., Somerville, I. D., Jones, N. S., Cleal, C. J., Collinson, J. D., Waters, R. A., Besly, B. M., Dean, M. T., Stephenson, M. H., Davies, J. R., Freshney, E. C., Jackson, D. I., Mitchell, W. I., Powell, J. H., Barclay, W. J., Browne, M. A. E., Leveridge, B. E., Long, S. L. & McLean, D. (Eds.), A Revised Correlation of Carboniferous Rocks in the British Isles, Special Report No. 26, The Geological Society, London, 49–56.

Devereux, P. & Nash, G., 2014, Indications of an acoustic landscape at Bryn Celli Ddu, Anglesey, North Wales, Time and Mind, 7(4), 385–390.

Edwards, N., 2013, *A Corpus of Early Medieval Inscribed Stones and Stone Sculptures in Wales: North Wales*, vol. 3, University of Wales Press, Cardiff, 544 pp.
Elis-Gruffydd, D., 2008, Owain Glyndwr, China and Trefor Quarry, *Welsh Stone Forum Newsletter*, 5, 2–3.
Fairbairn, W., 1849, *An Account of the Construction of the Britannia and Conway Tubular Bridges*, John Weale & Longmans, London, 291 pp.
Farr, G., Graham, J. & Stratford, C., 2014, Survey characterisation and condition assessment of Palustriella dominated springs 'H7220 Petrifying springs with tufa formation (Cratoneurion)' in Wales, Centre for Ecology and Hydrology and the British Geological Survey (NERC) for Natural Resources Wales, Evidence Report No. 136, 211 pp.
Fedden, R. & Thomson, J., 1957, *Crusader Castles*, John Murray, London, 127 pp.
Geldard, J. S., McNeal, J., Rhodes, S. J., Wiggett, A. & Williams, J. K., 2011, *Strategic Stone Study: A Building Stone Atlas of Lancashire*, 21 pp.
Ghey, E., Edwards, N. & Johnston, R., 2008, Categorizing roundhouse settlements in Wales: A critical perspective, *Studia Celtica*, 42, 1–25.
Gibbons, W., 1987, Menai strait fault system: An early Caledonian terrane boundary in north Wales, *Geology*, 15(8), 744–747.
Gibbons, W. & Ball, M. J., 1991, A discussion of the Monian Supergroup stratigraphy in NW Wales, *Journal of the Geological Society*, London, 148, 5–8. doi: 10.1144/gsjgs.148.1.0005.
Graham, J. & Farr, G., 2014, Petrifying springs in Wales, *Field Bryology*, 112, 19–29.
Greenly, E., 1919, *The Geology of Anglesey*, Memoir of the Geological Survey of Great Britain HMSO, London, 2 vols, 980 pp.
Greenly, E., 1928, The lower carboniferous rocks of the Menaian region of Carnarvonshire; their petrology, succession and physiography, *Quarterly Journal of the Geological Society of London*, 84, 382–439.
Greenly, E., 1932, The stones of the castles, Anglesey Antiquarian Society Transactions, 50–56.
Greenly, E., 1938, The red measures of the Menaian region of Carnarvonshire, *Quarterly Journal of the Geological Society of London*, 94, 331–345.
Greenly, E., 1944a, The Cambrian Rocks of Arvon, *Geological Magazine*, 81(5), 170–175.
Greenly, E., 1944b, The Ordovician Rocks of Arvon, *Quarterly Journal of the Geological Society of London*, 100, 75–83.
Greenly, E., 1945, The Arvonian Rocks of Arvon, *Quarterly Journal of the Geological Society of London*, 100, 269–284.
Greenly, E., 1946, The geology of the city of Bangor, *Proceedings of the Liverpool Geological Society*, 19, 105–112.
Gresham, C. A., 1938, The Roman Fort at Tomen-y-Mur, *Archaeologia Cambrensis*, 93, 192–211.

Gwyn, D. R., 2015, Welsh slate: The archaeology and history of an industry, Royal Commission of Ancient and Historic Monuments, Wales, Aberystwyth, 290 pp.

Gwyn, D. R. & Davidson, A., 1995, Gwynedd slate quarries: An archaeological survey 1994–5, Gwynedd Archaeological Trust report No. 152, 51 pp + Appendices & Plates.

Harris Jones, H., 2000, Chapter 16: Brickmaking in the Caernarfon area, in: Hughes, T. M., Mumford, M. R. & Ellis, M. R. (Eds.), *Caernarfon: The Millennium Town*, Mumford Books, Colwyn Bay, 120 pp.

Harris, C., Williams, G., Brabham, P., Eaton, G. & McCarroll, D., 1997, Glaciotectonized Quaternary sediments at Dinas Dinlle, Gwynedd, North Wales and their bearing on the style of deglaciation in the eastern Irish Sea, *Quaternary Science Reviews*, 16, 109–127.

Haslam, R., Orbach, J. & Voelcker, A., 2009, *The Buildings of Wales: Gwynedd, Pevsner Architectural Guides,* Yale University Press, New Haven & London, 789 pp.

Hassall, M. W. C. & Tomlin, R. S. O., 1977, Roman Britain in 1976: II inscriptions, *Britannia*, 8, 426–449.

Haycock, A., 2017, Halkyn Marble of Flintshire, *Welsh Stone Forum Newsletter*, 14, 3–6.

Haycock, A., 2018, Reinvestigation of Carboniferous sandstones, northeast Wales (Wrexham and Llangollen), *Welsh Stone Forum Newsletter*, 15, 3–6.

Hazzledine Warren, S., 1919, A stone-axe factory at Graig-Lwyd, Penmaenmawr, *The Journal of the Royal Anthropological Institute of Great Britain and Ireland*, 49, 342–365.

Hemp, W. J., 1931, The chambered cairn of Bryn Celli Ddu, *Archaeologia Cambrensis*, 86, 216–258.

Hemp, W. J., 1935, The chambered cairn known as Bryn yr Hen Bobl, near Plas Newydd, Anglesey, *Archaeologia*, 85, 253–292.

Hesselbo, S. P., Bjerrum, C. J., Hinnov, L. A., MacNiocaill, C., Miller, K. G., Riding, J. B., van de Schootbrugge, B. et al., 2013, Mochras borehole revisited: A new global standard for Early Jurassic earth history, *Scientific Drilling*, 16, 81–91.

Hopewell, D., 2018a, Roman Anglesey: Recent discoveries, *Britannia*, 49, 313–322.

Hopewell, D., 2018b, The Tre'r Ceiri conservation project: Re-examination of an iconic hillfort, Pre-publication Report no. 1417, Gwynedd Archaeological Trust.

Horák, J. M., 1993, The Late Precambrian Coedana and Sarn Complexes, Northwest Wales; a Geochemical and Petrological study. Unpublished Ph.D. Thesis, University of Wales, 415 pp.

Horák, J., 2005, Mona Marble: Characterisation and usage, in: Coulson, M. (Ed.), *Stone in Wales, Materials, Heritage and Conservation*, Cadw, Cardiff, 2–5.

Horák, J., 2011, New production of ornamental stones from the UK, *Welsh Stone Forum Newsletter*, 8, 27.

Horák, J., 2013, Geological sources and selection of stone, in: Edwards, N. (Ed.), *A Corpus of Early Medieval Inscribed Stones and Stone Sculptures in Wales: North Wales*, vol. 3, University of Wales Press, Cardiff, 30–40.

Horák, J., 2020, Building stone from Pwllheli, *Welsh Stone Forum Newsletter*, 17, 3–4.

Horák, J., Doig, R., Evans, J. A. & Gibbons, W., 1996, Avalonian magmatism and terrane linkage: New isotopic data from the Precambrian of North Wales, *Journal of the Geological Society, London*, 153, 91–99.

Hose, T. A., 1995, Selling the story of Britain's stone. *Environmental Interpretation*, 10, 16–17.

Hose, T. A., 2008, Towards a history of geotourism: Definitions, antecedents and the future, in: Burek, C. V. & Prosser, C. D. (Eds.), *The History of Geoconservation*, Geological Society, London, Special Publications, 300, 37–60.

Howe, J. A., 1910, *The Geology of Building Stones*, Edward Arnold, London, 455 pp.

Howells, M. F., 2007, *British Regional Geology: Wales*, British Geological Survey, Keyworth, Nottingham, 230 pp.

Howells, M. F. & Smith, M., 1997, The geology of the country around Snowdon, Memoir of the British Geological Survey, British Geological Survey & Her Majesty's Stationary Office (HMSO), London, 104 pp.

Howells, M. F., Reedman, A. J. & Leveridge, B. E., 1985, Geology of the country around Bangor: Explanation for 1:50,000 geological sheet 106 (England & Wales), Memoirs of the Geological Survey of Great Britain, England and Wales (sheet - new series), Her Majesty's Stationary Office (HMSO), London, 36 pp.

Hughes, P. D., 2009, Loch Lomond Stadial (Younger Dryas) glaciers and climate in Wales, *Geological Journal*, 44, 375–391.

Hughes, T., Horak, J., Lott, G. & Roberts, D., 2016, Cambrian age Welsh Slate: A global heritage stone resource from the United Kingdom, *Episodes*, 39(1), 45–51.

Hyslop, E. & Lott, G., 2007, Rock of ages: The story of British Granite, The Building Conservation Directory, http://www.buildingconservation.com/articles/rockofages/rockofages.htm.

Ixer, R. A., 2001, An assessment of copper mineralization from the Great Orme Mine, Llandudno, North Wales, as ore in the Bronze Age, *Proceedings of the Yorkshire Geological Society*, 53(3), 213–219.

Jackson, J., 1937, *Tacitus, Annals, Books XIII–XVI*, Loeb Classical Library, London and Cambridge, MA, 14.29–14.33.

Jenkins, D. A., 1995. Mynydd Parys Copper Mines. *Archaeology in Wales*, 35, 35–37.

Johnstone, N, 1999, Cae Llys, Rhosyr: A court of the Princes of Gwynedd, *Studia Celtica*, 33, 251–295.

Jones, R. M., 1982, *The North Wales Quarrymen 1874–1922*, University of Wales Press, Cardiff, 359 pp.

Jones, R. C., 1992, *Felinheli: A Personal History of the Port of Dinorwic*, Bridge Books, Dromore, Lisburn, 160 pp.

Jones, A. Ff., 2010, King Edward I's Castles in North Wales: Now and tomorrow, in: Williams, D. M. & Kenyon, J. R. (Eds.), *The Impact of the Edwardian Castles in Wales*, Oxbow Books, Oxford, United Kingdom, 198–202.

Jones, N. W. & Hankinson, R., 2016. Garth Quarry, Minffordd, Gwynedd—Heritage Impact Assessment. CPAT Report No. 1435, 91 pp.

Kaur, G., Singh, S., Ahuja, A. & Singh, N. D., 2020, *Natural Stone and World Heritage: Delhi-Agra, India*, Natural Stone and World Heritage 2, CRC Press, Taylor & Francis Group, London, UK, 177 pp.

Kenney, J., 2012, Assessment and scheduling enhancement Maenan Abbey, Llanrwst, Project No. G2213, Report No. 1039, Gwynedd Archaeological Trust, Bangor, 27 pp + 33 plates.

Kenney, J., 2017, Group VII Axe-working sites and stone sources, Llanfairfechan, Conwy, Report and gazetteer, Report No. 1416, Gwynedd Archaeological Trust, 63 pp.

Kenney, J. & Parry, L., 2012, Excavations at Ysgol yr Hendre, Llanbeblig, Caernarfon: A possible construction camp for Segontium fort and early medieval cemetery, *Archaeologia Cambrensis* 161, 249–284.

King, A., 2011, *Strategic Stone Study: A Building Stone Atlas of Cheshire*, English Heritage, Swindon, 18 pp.

Knoop, D. & Jones, C. P., 1932, Castle building at Beaumaris and Caernarvon in the early fourteenth century, *Transactions of the Quatuor Coronati Lodge*, 45, 4–47.

Kuenen, P. H., 1953, Graded bedding with observations on lower Palaeozoic rocks of Britain. *Verhandelingen der Koninklijke Nederlandse Akademie von Wetenschappen Afd. Natuurkunde.*, 20(3), 1–47.

Lapworth, C., 1879, On the tripartite classification of the lower Palaeozoic rocks. *Geological Magazine*, 6, 1–15.

Lewis, C. A., 1996, Prehistoric mining at the great Orme: Criteria for the identification of early mining, Unpublished M.Phil, University of Bangor, 184 pp.

Lilley, K. D., 2010, The landscapes of Edward's new towns: Their planning and design, in: Williams, D. M. & Kenyon, J. R. (Eds.), *The Impact of the Edwardian Castles in Wales*, Oxbow Books, Oxford, United Kingdom, 99–113.

Lloyd Jones, E., 2016, Rediscovering skilled workers and their legacy at Beaumaris Castle, *Anglesey Antiquarian Society Journal,* 2015–2016, 124–137.

Llwyd, A., 1833, A history of the Island of Mona, or Anglesey; being the prize essay to which was adjudged the first premium at the royal Beaumaris Eisteddfod 1832, R. Jones, & Longman, London, 474.

Longley, D., 2010, Gwynedd before and after the conquest, in: Williams, D. M. & Kenyon, J. R. (Eds.), *The Impact of the Edwardian Castles in Wales*, Oxbow Books, Oxford, United Kingdom, 16–26.

Lott, G., 2001, Geology and building stones in the East Midlands, *Mercian Geologist*, 15(2), 97–122.

Lott, G., 2009, The petrography of some Carboniferous sandstones from north-east Wales, *Welsh Stone Forum Newsletter*, 6, 10–12.

Lott, G., 2010, The building stones of the Edwardian Castles, in: Williams, D. M. & Kenyon, J. R. (Eds.), *The Impact of the Edwardian Castles in Wales*, Oxbow Books, London, 114–120.

Lott, G., 2012, Strategic stone study: A building stone atlas of West & South Yorkshire, English Heritage, 25 pp.

Lynch, F., 1991, *Prehistoric Anglesey*, 2nd Edition, Anglesey Antiquarian Society, Llangefni, 411 pp.

Malaws, B. A., 2006, National Language Centre, Nant Gwrtheyrn, RCAHMW; Coflein. https://coflein.gov.uk/en/site/33025/details/national-language-centre-nant-gwrtheyrn.

Malchow, H. L., 1992, Chapter 1: Festiniog to Westminster: Samuel Holland (1803–1892), in: *Gentleman Capitalists: The Social and Political World of the Victorian Businessman*, Stanford University Press, Palo Alto, CA, 18–77.

Manning, H. P., 2002, English money and Welsh rocks: Divisions of language and divisions of labor in nineteenth-century Welsh slate quarries, *Comparative Studies in Society and History*, 44(3), 481–510.

Marsden, J., 2009, *Penrhyn Castle*, The National Trust, Swindon, United Kingdom, 68 pp.

Nakamura, N. & Borradaile, G., 2001, Do reduction spots predate finite strain? A magnetic diagnosis of Cambrian slates in North Wales, *Tectonophysics*, 340, 133–139.

Nash, G., 2012, Report on later prehistoric portable art found at Crochan Caffo, Llangaffo, Gaerwen, Ynys Môn, *Archaeology in Wales*, 52, 133–135.

Nash, G., James, C. & Wellicome, T., 2014, The excavation of the Neolithic portal dolmen monument of Perthi Duon, Ynys Môn (Anglesey), *Archaeology in Wales*, 54, 25–34.

Neaverson, E., 1947, *Mediaeval Castles in North Wales: A Study of Sites, Water Supply and Building Stones*, University of Liverpool Press and Hodder & Stoughton, London, 54 pp.

Neuman, R. B. & Bates, D. E. B., 1978, Reassessment of Arenig and Llanvirn age (early Ordovician) brachiopods from Anglesey, North West Wales, *Palaeontology*, 21(3), 571–613.

Nevell, R., 2020, The archaeology of slighting: A methodological framework for interpreting castle destruction in the Middle Ages, *Archaeological Journal*, 177(1), 99–139.

Nichol, D., 2005, Geological provenance of Caernarfon Castle and town walls, in: Bassett, M. G., Deisler, V. K. & Nichol, D. (Eds.), *Urban Geology in Wales*, vol. 2, National Museum of Wales, Cardiff, 204–208.

O'Neil, B. H. St. J., 1944, Criccieth Castle, Caemarvonshire, *Archaeologia Cambrensis*, 98(1), 1–51.

Owen, E., 1917, The fate of the structures of Conway Abbey, and Bangor and Beaumaris Friaries, *Y Cymmrodor*, 27, 70–114.

Palmer, T., 2007, Egryn stone: A forgotten Welsh freestone, *Archaeologia Cambrensis*, 156, 149–160.

Parker-Pearson, M., Chamberlain, A., Mandy, J., Richards, M., Sheridan, A., Curtis, N., Evans, J., Gibson, A., Hutchison, M., Mahoney, P., Marshall, P., Montgomery, J., Needham, S., O'Mahoney, S., Pellegrini, M. & Wilkin, N., 2016, Bell beaker people in Britain: Migration, mobility and diet, *Antiquity*, 90(351), 620–637.

Parry, L. W. & Kenney, J., 2012. Archaeological discoveries along the Porthmadog bypass, *Archaeology in Wales*, 52, 113–132.

Pereira, D., 2019, *Natural Stone and World Heritage: Salamanca (Spain)*, CRC Press, Taylor & Francis Group, London, United Kingdom, 105 pp.

Pereira, D. & Van den Eynde, V. C., 2019, Heritage stones and Geoheritage, *Geoheritage*, 11, 1–2.

Phillips, C. W., 1932, The excavation of a hut site at Parc Dinmor, Penmon, Anglesey, *Archaeologia Cambrensis*, 87, 247–59.

Piers, C. R., 1916, Carnarvon Castle, *Transactions of the Honourable Society of Cymmrodorion*, Session 1915–1916, 1–74.

Prestwich, M., 2010, Edward I and wales, in: Williams, D. M. & Kenyon, J. R. (Eds), *The Impact of the Edwardian Castles in Wales*, Oxbow Books, Oxford, United Kingdom, 1–8.

Price, W. T. & Ronck, C. L., 2019, Quarrying for world heritage designation: Slate Tourism in North Wales, *Geoheritage*, 11, 1839–1854.

Pugh, W. J., 1923, The geology of the district around Corris and Aberllefenni (Merionethshire), *Quarterly Journal of the Geological Society*, 79(4), 508–545.

RCAHMW (undated), Cadw parks and gardens register text description of Plas Tan-y-bwlch Garden, Maentwrog. Parks and Gardens Register Number PGW(GD)031(GWY), 16 pp.

RCAHMW, 1937 (reprinted 1960), Anglesey: A survey and inventory by the royal commission on ancient and historical monuments in Wales and Monmouthshire, Her Majesty's Stationary Office, London, 589 pp.

RCAHMW, 1960, An inventory of the ancient monuments in Caernarvonshire. Volume III, Central, The Cantref of Arfon and the Commote of Eifionydd, The Royal Commission on Ancient and Historical Monuments in Wales and Monmouthshire, Her Majesty's Stationary Office, London, 287 pp.

Reedman, A. J., Leveridge, B. E. & Evans, R. B., 1984, The Arfon Group (Arvonian) of North Wales, *Proceedings of the Geologists' Association*, 95, 313–321.

Reynolds, F. F., Griffiths, S., Edwards, B. & Stanford, A., 2016, Bryn celli Ddu: Exploring a hidden ritual landscape, *Current Archaeology*, 310, 20–24.

Rhydderch-Dart, D., 2017, 'The great relief of the overcrowded slums': The development of Twthill in late 19th century Caernarfon, *Welsh History Review/Cylchgrawn Hanes Cymru*, 28(3), 515–548.

Richards, A. J., 2006, *Welsh Slate Craft: A Brief Account of Slate Working in Wales*, Gwasg Carreg Gwalch, Llanrwst, 104 pp.

Richards, A. J., 2007, *Gazeteer of Slate Quarrying in Wales*, Llygad Gwalch, Pwllheli, 351 pp.

Ritchie, M., 2018, A brief introduction to iron age settlement in Wales, *Internet Archaeology*, 48. doi: 10.11141/ia.48.2.

Roberts, B., 1979, *The Geology of Snowdonia and Llyn: An Outline and Field Guide*. Adam Hilger Ltd., Bristol, 183 pp.

Roberts, J., 2007, Seiont brickworks, Caernarfon; archaeological watching brief, Report 687, Gwynedd Archaeological Trust, Bangor, 3 pp.

Roberts, R., 2009, Upper carboniferous sandstones of North-East Wales, *Welsh Stone Forum Newsletter*, 6, 6–10.

Rushton, A. W. A. & Howells, M. F., 1998, Stratigraphical framework for the Ordovician of Snowdonia and the Lleyn Peninsula, British Geological Survey Research Report, RR/99/08, 38 pp.

Schofield, D. I., Evans, J. A., Millar, I. L., Wilby, P. R. & Aspden, J. A., 2008, New U–Pb and Rb–Sr constraints on pre-Acadian tectonism in North Wales, *Journal of the Geological Society, London*, 165, 891–894.

Schofield, D. I., Leslie, A. G., Wilby, P. R., Dartnall, R., Waldron, J. W. F. & Kendall, R. S., 2020, Tectonic evolution of Anglesey and adjacent mainland North Wales, in: Murphy, J. B., Strachan, R. A. & Quesada, C. (Eds), *Pannotia to Pangaea: Neoproterozoic and Paleozoic Orogenic Cycles in the Circum-Atlantic Region*. Geological Society, London, Special Publications, 503, 20 pp. doi: 10.1144/SP503-2020-9.

Sedgwick, A. & Murchison, R. I., 1835, On the Silurian and Cambrian Systems, exhibiting the order in which the older sedimentary strata succeed each other in England and Wales, Report of the British Association for the Advancement of Science Transactions, Dublin, 59–61.

Shipton, J., 2009, Field meeting reports: Anglesey. *Welsh Stone Forum Newsletter*, 6, 12–15.

Shipton, J., 2019, Flintshire & Denbighshire, *Welsh Stone Forum Newsletter*, 16, 27–31.

Shipton, J., 2020, Field trip, North West Wales. *Welsh Stone Forum Newsletter*, 17, 13–18.

Siddall, R., 2014, Jeremiah in slate valley: Taconic slate, or piment, Wordpress. https://orpiment.wordpress.com/2014/12/11/jeremiah-in-slate-valley-taconic-slate/.

Siddall, R., 2019, London pavement geology: No longer just a building stone resource for Londoners, *Welsh Stone Forum Newsletter*, 16, 13–14.

Siddall, R., Schroder, J. K. & Hamilton, L., 2016, A tour in three parts of the building stones used in the city centre; Part 1: From the Town Hall to the Cathedral, 17 pp. https://www.ucl.ac.uk/~ucfbrxs/Homepage/walks/Birmingham1-Centre.pdf.

Sims-Williams, P., 2004, Beuno [St Beuno], Oxford Dictionary of National Biography. doi: 10.1093/ref:odnb/2317.

Skempton, A., Chrimes, M., Cox, R. C., Cross-Rudkin, P. S. M., Rennison, R. W. & Ruddock, E. C. (Eds.), 2002. *A Biographical Dictionary of Civil Engineers in Great Britain and Ireland*, vol. 1, The Institute of Civil Engineers & Thomas Telford Ltd., London, 1500–1830.

Smith, A. G., 2005, The north-west Wales lithic scatters project, *The Journal of the Lithic Studies Society*, 26, 38–56.

Smith, M., 2007, Ffestiniog granite quarry, Chapter 6: Wales and adjacent areas, in: Stephenson, D. (Ed.), *Caledonian Igneous Rocks of Great Britain*, vol. 17, Geological Conservation Review, Joint Nature Conservation Committee (JNCC), UK, 4 pp.

Smith, G., 2018, Hillforts and Hut Groups of north-west Wales, *Internet Archaeology*, 48. doi: 10.11141/ia.48.6.

Snowdonia National Park, 2014, Camau daearegol Darwin trwy Eryri: y stori tu ôl I'r wal/Darwin's Geological footsteps through Snowdonia—the story behind the wall. Snowdonia National Park, National Trust &Natural Resources Wales. *Leaflet.* https://www.snowdonia.gov.wales/visiting/ogwen/darwins-snowdonia/darwins-wall.

Southern, D. W., 1995, *Bala Junction to Blaenau Ffestiniog*. Scenes from the Past, Railways of North Wales, No. 25, Foxline Publishing, Stockport, 112 pp.

Speed, J., 1611a, Anglesey Antiently Called Mona. Described 1610, National Library of Wales, http://hdl.handle.net/10107/1445590.

Speed, J., 1611b, Carnarvon both shyre and shire-towne with the ancient citie of Bangor described. Anno Domini 1610, National Library of Wales, http://hdl.handle.net/10107/1445798.

Speed, J., 1611c, Merionethshire described 1610. National Library of Wales, http://hdl.handle.net/10107/1445694.

Strachan, R. A., Collins, A. S., Buchan, C., Nance, R. D., Murphy, J. B. & D'Lemos, R. S., 2007, Terrane analysis along a Neoproterozoic active margin of Gondwana: Insights from U-Pb zircon geochronology. *Journal of the Geological Society of London*, 164, 57–60.

Swallow, R., 2019, Living the dream: The legend, lady and landscape of Caernarfon Castle, Gwynedd, North Wales, *Archaeologia Cambrensis*, 168, 153–195.

Talbot, J. & Cosgrove, J., 2011, *The Roadside Geology of Wales, Geologists' Association Guides No. 69*, Geologists' Association, London, 214 pp.

Taylor, A. J., 1950, Master James of St. George, *English Historical Review*, 65, 433–457.

Taylor, A. J., 1987, The Beaumaris Castle building account of 1295–1298, in: Kenyon, J. R. & Avent, R. (Eds.), *Castles in Wales and the Marches: Essays in Honour of D. J. Cathcart King*, University of Wales Press, Cardiff, 125–142.

Taylor, A. J., 1993, *Caernarfon Castle*, Cadw, Cardiff, 48 pp.

Taylor, A. J., 1995, The town and castle of Conwy: Preservation and interpretation, *The Antiquaries Journal*, 75, 339–363.

Taylor, A. J., 2015, *Caernarfon Castle*, Cadw, Cardiff, 44 pp.

Tellier, G., 2018, Neolithic and bronze age funerary and ritual practices in Wales, 3600–1200 BC, B642, BAR Publishing, 214 pp.

Thackray, J. C., 1976, The Murchison-Sedgwick controversy, *Journal of the Geological Society of London*, 132, 367–372.

Thomas, I. A., 2014, Quarrying industry in Wales: A history|Y diwydiant Chwareli yng Nghymru—hanes, The National Stone Centre, 224 pp.

Timberlake, S. & Marshall, P., 2018, Chapter 29, Copper mining and smelting in the British Bronze Age: New evidence of mine sites including some re-analysis of dates and ore sources, in: Ben-Yosef, E. (Ed.), *Mining for Ancient Copper Essays in Memory of Beno Rothenberg*, Eisenbrauns & Winona Lake, Indiana and Emery and Claire Yass Publications in Archaeology, Tel Aviv University, Tel Aviv, 418–431.

Tremlett, W. E., 1962, The geology of the Nefyn-Llanaelhaiarn area of North Wales, *Liverpool and Manchester Geological Journal*, 3(1), 157–176.

Tremlett, W. E., 1997, Geochemical variation in the Penmaenmawr intrusion (North Wales), *Geological Journal*, 32, 173–187.

Tucker, R. D. & Pharoah, T. C., 1991, U-Pb zircon ages for Late Precambrian igneous rocks in southern Britain, *Journal of the Geological Society*, London, 148, 435–443.

Turner, R., 2008, *Plas Mawr, Conwy*, Cadw, Cardiff, 44 pp.

Turner, R., 2010, The life and career of Richard the Engineer, in: Williams, D. M. & Kenyon, J. R. (Eds.), *The Impact of the Edwardian Castles in Wales*, Oxbow Books, Oxford, United Kingdom, 46–58.

UNESCO, 2019, Operational Guidelines for the Implementation of the World Heritage Convention, United Nations Educational, Scientific and Cultural Organization (UNESCO), Intergovernmental Committee for the Protection of the World Cultural and Natural Heritage, World Heritage Convention, WHC.19/01, 177 pp.

Warren, P. T., Price, D., Nutt, M. J. C. & Smith E. G., 1984, Geology of the country around Rhyl and Denbigh, Memoir of the British Geological Survey, sheets 95 and 107 (England and Wales).

Waters, C. N., Browne, M. A. E, Dean, M. T. & Powell, J. H., 2007, A lithostratigraphical framework for Carboniferous successions of Great Britain (Onshore). British Geological Survey Research Report, RR/07/01, 60pp.

Watson, J., 1911, *British and Foreign Building Stones: A Descriptive Catalogue of the Specimens in the Sedgwick Museum*, Cambridge University Press, Cambridge, MA, 483 pp.

Watson, J., 1916, *British and Foreign Marbles and other Ornamental Stones: A Descriptive Catalogue of the Specimens in the Sedgwick Museum*, Cambridge University Press, Cambridge, MA, 485 pp.

Welsh Government, 2020, Welsh Government Press Release: Welsh Slate Landscape nominated for UNESCO World Heritage status, 24 January 2020. https://www.gov.uk/government/news/welsh-slate-landscape-nominated-for-unesco-world-heritage-status.

Wheatley, A., 2010, Caernarfon Castle and its mythology, in: Williams, D. M. & Kenyon, J. R. (Eds.), *The Impact of the Edwardian Castles in Wales*, Oxbow Books, Oxford, United Kingdom, 129–139.

Wheeler, R. E. M., 1924, *Segontium and the Roman Occupation of Wales*, The Honourable Society of Cymmrodorion, London, 186 pp.

White, R. B., 1985, Excavations in Caernarfon 1976–77, *Archaeologica Cambrensis*, 134, 53–105.

Williams, M., 1991, *The Slate Industry*, Shire Publications Ltd., Oxford, 32 pp.

Williams, R. A., 2018, The great Orme bronze age copper mine: Linking ores to metals by developing a geochemically and isotopically defined mine-based methodology, in: Montero-Ruiz, I. & Perea, A. (Eds.), *Archaeometallurgy in Europe IV, Bibliotheca Prehistorica Hispana*, vol. XXXIII, Consejo Superior de Investigaciones Científicas, Instituto de Historia, Madrid, 29–47.

Williams, J. Ll. W. & Jenkins, D. A., 1999, A petrographic investigation of a corpus of bronze age cinerary urns from the Isle of Anglesey, *Proceedings of the Prehistoric Society*, 65, 189–230.

Williams, D. M. & Kenyon, J. R., 2010, *The Impact of the Edwardian Castles in Wales*, Oxbow Books, Oxford, United Kingdom, 240 pp.

Williams, J. Ll. W., Kenney, J. & Edmonds, M., 2011, Graig Lwyd (Group VII) assemblages from Parc Bryn Cegin, Llandegai, Gwynedd, Wales: Analysis and interpretation, in: Davis, V. & Edmonds, M. (Eds.), *Stone Axe Studies III*, Oxbow Books, Oxford & Oakville, 261–278.

Young, T. P. & Gibbons, W., 2007, Llanbedrog, Chapter 6: Wales and adjacent areas, in: *Caledonian Igneous Rocks of Great Britain*, vol. 17, Geological Conservation Review, Joint Nature Conservation Committee (JNCC), UK, 2 pp.

Young, T. P., Gibbons, W. & McCarroll, D., 2002, Geology of the Country around Pwllheli, Memoir of the British Geological Survey, Sheet 134 (England a7 Wales), British Geological Survey, Keyworth, 136 pp.

WEB RESOURCES

Legacies of British Slave-Ownership, UCL History, Accessed 6 August 2020, https://www.ucl.ac.uk/lbs/.

London Pavement Geology, http://londonpavementgeology.co.uk/.

UNESCO Geopark Geomôn, http://www.unesco.org/new/en/natural-sciences/environment/earth-sciences/unesco-global-geoparks/list-of-unesco-global-geoparks/united-kingdom/geomon/.

UNESCO, 1986, World Heritage Centre; The Castles and Town Walls of King Edward in Gwynedd, https://whc.unesco.org/en/list/374.

Welsh Stone Forum, National Museum of Wales / Fforwm Cerrig Gymru, Amgueddfa Genedlaethol Cymru, https://museum.wales/curatorial/geology/welsh-stone-forum/.

Index

ashlar masonry 32, 49, 61, 68, 70,
75, 103, 106, 117, 119, 130, 145,
192–194, 196, 207, 208, 213, 228,
233, 235, 237, 238, 244, 247,
254–256, 258, 271, 272
Ashmolean Museum 114
Assheton-Smith Family 25, 207,
209, 210, 263
augen schists 168, 228, 238
Augustinian Order 45, 121
Australia xiii, 27, 80
Australia Quarry 25
axes *see* Lithic tools and stone axes

Bachwen Dolmen 21, 87, 277
Bala 58, 60
ballast 78–80, 162–163, 231–232,
246, 249, 250, 262, 263; *see also*
Railway Ballast
Bangor xi, 20–22, 24, 42, 47–50, 56,
57, 72, 74, 76–79, 112, 117, 140,
147, 170, 200, 202, 206, 213, 219,
226–232
Bangor Cathedral 46, 200,
226–229, 277
Bangor Formation 21–22, 229
Bangor Friary 147
Bangor Mountain 21
Bangor Police Station 231
Bangor Post Office 76, 230
Bangor University 21, 69, 72, 226,
227, 229–231
barbican 189
Bardsey Island 112, 202
Barmouth 176, 275
Barmouth Formation 28, 90
Barnwell, Reverend E. L. 107
Baron Hill 234
Baron Hill Beds 21, 36
Baron Penrhyn of Llandegai 24
Baron Willoughby de Eresby 125
basalt 18, 21, 232, 256
Basement Beds 42–46, 54, 70, 74,
85, 89, 101, 103–105, 110, 111,
113, 117–121, 130, 151, 155–158,
160, 166–170, 192–194, 204–206,
208–209, 228, 233–234, 238,
240–243, 246, 249, 252, 256, 257,
263, 210, 270–271, 273

Basingwerk Abbey 71
Bath Stone 75–76, 210, 261
Bavaria 76
Bayly, Nicholas Sir. 210
Baynes, E. Neil 87, 109
Beaker People 89
beam engine 26
Beaumaris xiv, 1, 2, 7, 18, 21, 35,
36, 42, 48, 72, 77, 187–196, 199,
200, 206, 232–239, 257, 272
Beaumaris Castle xi, xii, 1, 2, 4, 7,
15, 18, 44, 45, 49, 133, 135, 136,
140–142, 153, 160, 187–196, 266,
270, 271
Beaumaris Castle Gatehouse 189
Beaumaris Gaol 233, 235
Beddgelert 115, 218
Beeston Castle 74, 129
Belfast 80, 163
Bendigeidfran the Giant 171
Benllech 45, 46, 48, 87, 192,
225, 228
Benllech Cromlech 87, 277
Benllech Sandstone 43
Berw Shear Zone 17, 86
Berwig Stone 71, 161
Berwyn Slate 40
Bethesda 22, 24, 26, 27, 207,
223, 232
Bettisfield Formation 72
Betws-y-Coed 37, 76, 218
Beynardus 139
Birdlip Limestone Formation 215
Birmingham 61, 70, 272
Birmingham Town Hall 48,
235–236, 272
Bishop of Gwynedd 226
bituminous limestone 48, 216
black marble 48, 216
blacksmith 141, 188
Blaenau Ffestiniog xiv, 1, 2, 5,
32–35, 58–59, 61, 70, 124,
212, 276
Blaenau Ffestiniog Slate 34–35, 37,
39, 224, 225, 259–262, 264–266
blondins 26
Blue Pennant Sandstone 72–73,
226, 250, 255, 259
blueschist 17–19, 85, 86, 237, 266

Cwm Eigiau Formation 37, 125
Cwm Eigiau Quarry 38, 278
Cwm-y-Glo 25
Cwt-y-Bugail Quarry 33, 35, 225, 278

D-plan tower 129, 166, 167
Dafydd ap Gruffudd 131, 134
dais 149, 153
Darbishire Family 57, 62
Darwin, Charles 6, 13
Darwin's Wall 6, 28
Davies, B. Price 230
Davies, Ivor 56
Davies, R. Trefor 62
Daviesiella llangollensis 46, 103, 241
Defferd, James 210
Deganwy 202, 220, 254
Deganwy Castle 37, 124, 127–129, 187, 196
Deiniol Shopping Centre 50
Deiniolen 25
Delamere Forest 160
Delamere Sandstone Member 74, 76, 105, 158, 160
Denbigh Grits Formation 40, 99, 202, 256
Deva Victrix 74, 97
Devonian System 23, 39–41, 46, 110
diamicton 67, 68
Diffwys Quarry 33, 35, 278
Din Lligwy 109, 111, 117, 272, 278
Dinas (Criccieth) 64, 129, 130
Dinas Dinlle 63, 67, 95, 213, 278
Dinas Dinorwig Hillfort 14, 95, 278
Dinas Gynfor Hillfort 94, 278
Dinas Quarry 58
Dinmoor Quarry 48
Dinorben Quarry 48, 216, 278
Dinorwic Fault 20, 208, 219
Dinorwic Formation 47
Dinorwig Hydroelectric Power Station 25
Dinorwig Quarries 5, 14, 22, 24–25, 33, 35, 207, 278
Doc Fictoria 44, 78, 165, 219, 246, 247, 250, 278
Dol Ifan Githin Quarry 30, 278
Dol-cyn-afon Formation 30, 31, 58, 130, 260, 261
Dolbadarn Castle 124, 126–127, 278
Dolbenmaen 206
dolerite 11, 54, 55, 58–60, 89, 107, 165, 213, 225, 247, 260, 261, 262
Dolgarrog Volcanic Formation 38, 256
Dolgellau 14
Dolgellau Formation 29, 30
Dolmen 83, 85–87
Dolwyddelan 2
Dolwyddelan Castle 4, 53, 124–126, 278
Dolwyddelan Slate 37
Dolwyddelan Syncline 37
Dorothea Quarry 26, 278
Douglas, Edward Gordon 24
Druid's Circle 90, 278
Drummond-Burrell, Peter 125
dry-stone walls 2, 18, 21, 36, 56, 82, 95, 150, 155
Dublin 26, 58, 80, 163, 227, 250
Duke of Lancaster 200
Duke of Wellington 210
Dunoding xiv, 123
Dyffryn Ardudwy 83

earthworks 88, 98, 99, 100, 107, 108, 112, 123, 124, 256
Eben Fardd 266; *see also* Thomas, Ebenezer
Eccleston 74
Edward of Caernarfon *see* King Edward II
Edward the Black Prince 184
Edwen Sandstone 43
Egryn Stone 28–29, 119, 174, 176–177, 197, 203–205, 259, 275, 278
Eifionydd 111, 129
Eisteddfod 236
Eleanor of Castile 134, 144, 146
Eleanor of England 141
Eleanor of Provence 134
Elidir Fawr 24, 25
Elland Flags 72
Emmota 141
enamelled slate 223
Enderby and Stoney Stanton Granite Company 64